RECOGNIZING AND RESPONDING TO NORMALIZATION OF DEVIANCE

RECOGNIZING AND RESPONDING TO NORMALIZATION OF DEVIANCE

CENTER FOR CHEMICAL PROCESS SAFETY
OF THE
AMERICAN INSTITUTE OF CHEMICAL ENGINEERS
New York, NY

WILEY

This edition first published 2018

© 2018 the American Institute of Chemical Engineers

A Joint Publication of the American Institute of Chemical Engineers and John Wiley & Sons, Inc.

Registered Office
John Wiley & Sons, Inc., 111 River Street, Hoboken, NJ 07030, USA

Editorial Office
111 River Street, Hoboken, NJ 07030, USA

For details of our global editorial offices, customer services, and more information about Wiley products visit us at www.wiley.com.

Wiley also publishes its books in a variety of electronic formats and by print-on-demand. Some content that appears in standard print versions of this book may not be available in other formats.

Library of Congress Cataloging-in-Publication Data is available.

ISBN: 9781119506713

Cover Design: Wiley
Cover images: (top left): Photograph of Bhopal on cover provided by Dennis Hendershot, © 2004; (top right) Permission granted by CCPS for use of cover image adapted from Figure 3.1. All rights reserved; (bottom left) © US Chemical Safety Board. Public Domain; (bottom right and background) © ASSOCIATED PRESS/AP Images; (background): © Jasmin Merdan/Getty Images

Printed in the United States of America
V10003988_082318

In Memoriam

Robert (Bob) Walters

1957 – 2016

This book is dedicated to Bob Walters, whose organization skills, quick wit, access to notable quotes and technical knowledge made him a good collaborator for *Recognizing and Responding to Normalization of Deviance*. As President of AntiEntropics, Inc., Bob provided expert guidance and writing for a number of CCPS Books and *The Business Case for Process Safety*. I worked with him on several including *Recognizing Catastrophic Incident Warning Signs in the Process Industries*, which is now a top seller for CCPS. We will miss Bob and the contributions that he made to CCPS. He will be remembered fondly.

Louisa A. Nara
CCPS Global Technical Director

TABLE OF CONTENTS

LIST OF TABLES

LIST OF FIGURES

GLOSSARY

Note: This Glossary contains the terms specific to this book and process safety related terms from the CCPS Process Safety Glossary. The specific CCPS process safety related terms in this book are current at the time of publication; please access the CCPS website for potential updates to the CCPS Glossary. www.aiche.org/ccps

Abnormal Situation—Disturbance in an industrial process with which the basic process control system of the process cannot cope. In the context of hazard evaluation procedures, synonymous with deviation.

Acceptable Risk—The average rate of loss that is considered tolerable for a given activity.

Accident—An incident that results in significant human loss.

Administrative Control—Procedures that will hold human and/or equipment performance within established limits.

Alarm Management—Procedures, schematic, software, maintenance, documentation, hardware, logic, prioritization, characterization, etc., pertaining to the management of the process alarm system.

Asset Integrity—The condition of an asset that is properly designed and installed in accordance with specifications and remains fit for purpose.

Asset Integrity Management—A process safety management system for ensuring the integrity of assets throughout their life cycle.

Audit—A systematic, independent review to verify conformance with prescribed standards of care using a well-defined review process to ensure consistency and to allow the auditor to reach defensible conclusions.

Availability Heuristic—An availability heuristic is a mental shortcut that people build from their most immediate or recent examples that quickly come to mind when making a decision or evaluating a situation.

Baseline Risk Assessment—A process to characterize the current and potential threats to human health and the environment that may be posed by contaminants migrating to groundwater or surface water; releasing to air; leaching through soil; remaining in the soil and bio-accumulating in the food chain. The primary purpose of the baseline risk assessment is to provide risk managers with an understanding of the actual and potential risks to human health and the environment posed by the site and any uncertainties associated with the assessment. This information may be useful in determining whether a current or potential threat to human health or the environment warrants remedial action.

Catastrophic—A loss with major consequences and unacceptable lasting effects, usually involving significant harm to humans, substantial damage to the environment, and/or loss of community trust with possible loss of franchise to operate.

Configuration Management—The systematic application of management policies, procedures, and practices to assess and control changes to the hardware and software of a system and to maintain traceability of the configuration to the design basis throughout the system life. Configuration management is a specialized form of management of change.

Consequence(s)—The direct, undesirable result of an accident/incident sequence usually involving a fire, explosion, or release of toxic material. Consequence descriptions may be qualitative or quantitative estimates of the effects of an accident/incident.

Constrained Optimization—Optimizing an objective function with respect to some variables in the presence of constraints on those variables. Humans unceasingly rationalize situations based upon constrained optimization.

Controlled Document—Documents covered under a revision control process to ensure that up-to-date documents are available and out-of-date documents are removed from circulation.

Core Value—A value that has been promoted to an ethical imperative, accompanied with a strong individual and group intolerance for poor performance or violations of standards for activities that impact the core value.

Corrective Maintenance—Maintenance performed to repair a detected fault.

Critical Equipment—Equipment, instrumentation, controls, or systems whose malfunction or failure would likely result in a catastrophic release of highly hazardous chemicals, or whose proper operation is required to mitigate the consequences of such release. (Examples are most safety systems, such as area Lower Explosive Limit (LEL) monitors, fire protection systems such as deluge or underground systems, and key operational equipment usually handling high pressures or large volumes.)

Deviation - Disturbance in an industrial process with which the basic process control system of the process cannot cope. In the context of hazard evaluation procedures, synonymous with abnormal situation.

Effectiveness—The combination of process safety management performance and process safety management efficiency. An effective process safety management program produces the required work products of sufficient quality while consuming the minimum amount of resources.

Emergency Response Plan—A written plan which addresses actions to take in case of plant fire, explosion, or chemical release

Good Engineering Practices—Engineering, operating, or maintenance activities based upon established codes, standards, published technical reports, or recommended practices.

Hazards and Operability Study (HAZOP)— A systematic qualitative technique to identify process hazards and potential operating problems using a series of guide words to study process deviations. A HAZOP is used to question every part of a process to discover what deviations from the intention of the design can occur and what their causes and consequences may be. This is done systematically by applying suitable guidewords.

Heuristics—Any approach to problem solving, learning, or discovery that employs a practical method not guaranteed to be optimal or perfect, but sufficient for the immediate goals.

Incident Investigation—A systematic approach for determining the causes of an incident and developing recommendations that address the causes to help prevent or mitigate future incidents.

Incident Investigation Management System—A written document that defines the roles, responsibilities, protocols, and specific activities to be carried out by personnel performing an incident investigation.

Incident Investigation Team—A group of qualified people that examine an incident in a manner that is timely, objective, systematic, and technically sound to determine that factual information pertaining to the event is documented, probable cause(s) are ascertained, and complete technical understanding of such an event is achieved.

Incident Warning Sign—An indicator of a subtle problem that could lead to an incident.

Institutionalization—The organization exposes newcomers to deviant behaviors, often performed by authority figures, and explains those behaviors as organizationally normative

Intermediates—Materials from a process that are not yet completely finished product. They may be a mixture or compound.

Job Task Analysis—The analysis phase of the instructional systems design (ISD) model consists of a job task analysis based upon the equipment, operations, tools, and materials to be used as well as the knowledge and skills and attitudes required for each job position.

Layer of Protection Analysis (LOPA)—An approach that analyzes one incident scenario (cause-consequence pair) at a time, using predefined values for the initiating event frequency, independent protection layer failure probabilities, and consequence severity, in order to compare a scenario risk estimate to risk criteria for determining where additional risk reduction or more detailed analysis is needed. Scenarios are identified elsewhere, typically using a scenario-based hazard evaluation procedure such as a HAZOP Study.

Management of Change—A management system to identify, review, and approve all modifications to equipment, procedures, raw materials, and processing conditions, other than replacement in kind, prior to implementation to help ensure that changes to processes are properly analyzed (for example, for potential adverse impacts), documented, and communicated to employees affected.

Mechanical Integrity—A management system focused on ensuring that equipment is designed, installed, and maintained to perform the desired function. See Asset Integrity.

Mechanical Integrity Program—A program to ensure that process equipment and systems are and remain mechanically suitable for operation. It involves inspection, testing, upgrading and repairs of equipment, as well as written procedures to maintain on-going integrity of equipment. See Asset Integrity Management.

Normal Operation—Any process operations intended to be performed between startup and shutdown to support continued operation within safe upper and lower operating limits.

Normalization of Deviance—A gradual erosion of standards of performance because of increased tolerance of nonconformance. Also Normalization of Deviation.

Operating Procedures—Written, step by step instructions and information necessary to operate equipment, compiled in one document including operating instructions, process descriptions, operating limits, chemical hazards, and safety equipment requirements.

Performance Measure—A metric used to monitor or evaluate the operation of a program activity or management system.

Personal Protective Equipment (PPE)—Equipment designed to protect employees from serious workplace injuries or illnesses resulting from contact with chemical, radiological, physical, electrical, mechanical, or other workplace hazards. Besides face shields, safety glasses, hard hats, and safety shoes, PPE includes a variety of devices and garments such as goggles, coveralls, gloves, vests, earplugs, and respirators.

Pre-Startup Safety Review (PSSR)—A systematic and thorough check of a process prior to the introduction of a highly hazardous chemical to a process. The PSSR must confirm the following: Construction and equipment are in accordance with design specifications; Safety, operating, maintenance, and emergency procedures are in place and are adequate; A process hazard analysis has been performed for new facilities and recommendations have been resolved or implemented before startup, and modified facilities meet the management of change requirements; and training of each employee involved in operating a process has been completed.

Preventive Maintenance— Maintenance that seeks to reduce the frequency and severity of unplanned shutdowns by establishing a fixed schedule of routine inspection and repairs.

Process Flow Diagram (PFD)—A diagram that shows the material flow from one piece of equipment to the other in a process. It usually provides information about the pressure, temperature, composition, and flow rate of the various streams, heat duties of exchangers, and other such information pertaining to understanding and conceptualizing the process.

Process Hazard Analysis (PHA)—An organized effort to identify and evaluate hazards associated with processes and operations to enable their control. This review normally involves the use of qualitative techniques to identify and assess the significance of hazards. Conclusions and appropriate recommendations are developed. Occasionally, quantitative methods are used to help prioritized risk reduction.

Process Safety Information (PSI)—Physical, chemical, and toxicological information related to the chemicals, process, and equipment. It is used to document the configuration of a process, its characteristics, its limitations, and as data for process hazard analyses.

Rationalization—A mindset that enables system operators to convince themselves that their deviances are not only legitimate, but acceptable and perhaps even necessary.

Replacement in Kind—A replacement that satisfies the design specifications.

Risk—A measure of human injury, environmental damage, or economic loss in terms of both the incident likelihood and the magnitude of the loss or injury. A simplified version of this relationship expresses risk as the product of the likelihood and the consequences (i.e., Risk = Consequence x Likelihood) of an incident.

Risk Analysis—The estimation of scenario, process, facility and/or organizational risk by identifying potential incident scenarios, then evaluating and combining the expected frequency and impact of each scenario having a consequence of concern, then summing the scenario risks if necessary to obtain the total risk estimate for the level at which the risk analysis is being performed.

Risk Based Process Safety—The CCPS's process safety management system approach that uses risk-based strategies and implementation tactics that are commensurate with the risk-based need for process safety activities, availability of resources, and existing process safety culture to design, correct, and improve process safety management activities.

Risk Factor—Along with the probability that an event will occur (risk) risk factors are those factors of behavior, lifestyle, environment, or heredity associated with increasing or decreasing that probability.

Safety Instrumented System (SIS)—The instrumentation, controls, and interlocks provided for safe operation of the process.

Socialization—When behavior is often mediated by a system of rewards and punishments, aimed at determining whether the newcomer will or will not join the group by adopting the group's deviant behaviors.

Verification Activity— A test, field observation, or other activity used to ensure that personnel have acquired necessary skills and knowledge following training.

Worst-Case Scenario (WCS)—A release involving a hazardous material that would result in the worst (most severe) off-site consequences.

ACRONYMS AND ABBREVIATIONS

AIChE—American Institute of Chemical Engineers

CCPS—Center for Chemical Process Safety

CSB—Chemical Safety Board (U.S.)

HAZOP—Hazard and Operability Study

HIRA- Hazard Identification and Risk Analysis

HRO—High Reliability Organization

ISO—International Organization for Standardization

KSA—Knowledge, Skills, and Attitudes

MOC—Management of Change

NBIC—National Board Inspection Code

NFPA—National Fire Protection Association (U.S.)

PHA—Process Hazard Analysis

PPE—Personal Protective Equipment

PSI—Process Safety Information

PSM—Process Safety Management

PSSR—Pre-Startup Safety Review

PSV—Pressure Safety Valve

RBPS—Risk Based Process Safety

RMP—Risk Management Program (U.S. EPA)

FILES ON THE WEB

Access the documents accompanying Recognizing and Responding to Normalization of Deviance using a web browser at the following URL:

www.aiche.org/ccps/resources/Norm_Devtn

The password for the file is **Norm_dev2017**.

ACKNOWLEDGEMENTS

The American Institute of Chemical Engineers (AIChE) wishes to thank the Center for Chemical Process Safety (CCPS) and those involved in its operation, including its many sponsors whose funding made this project possible, and the members of the Technical Steering Committee, who conceived of and supported this book project. The members of the normalization of deviation subcommittee who worked with AntiEntropics, Inc. to produce this text deserve special recognition for their dedicated efforts, technical contributions, and overall enthusiasm for creating a useful addition to the process safety concept book series. CCPS also wishes to thank the subcommittee members' respective companies for supporting their involvement in this project.

The chairperson of the normalization of deviation subcommittee was Jennifer Mize of Eastman Chemical Company. The CCPS staff liaison was Dan Sliva. The members of the CCPS guideline subcommittee were:

- Steve Arendt – ABS Group
- Michelle Brown –FMC Corporation
- James Caudill - Marathon Petroleum Company LP
- Kelly-Ann Charles – Methanex Corporation
- Joey Cranston – Albemarle Corporation
- Scott Haney – Marathon Petroleum Company LP
- David Hill – Occidental Chemical Corporation (OxyChem)
- Greg Horton – SABIC
- Patti Jones – Praxair, Incorporated
- Jai Karia – Chevron Corporation
- Neil Maxson – Covestro
- Mikelle Moore – Buckman International, Incorporated
- Louisa Nara – CCPS
- Mark Paradies - System Improvements Incorporated
- Michael Pelupessy – Akzo Nobel
- Sara Saxena - BP
- Joan Schork – Baker Engineering and Risk Consultants, Inc.
- Adrian Sepeda – AIChE Emeritus
- Tony Strawhun – Afton Chemical Corporation
- Ken Tague – Archer Daniels Midland Co.
- Karen Tancredi - Chevron
- Daniel Wilczynski - Marathon Petroleum Company LP
- Elliot Wolf - Syngenta

Robert J. Walter of AntiEntropics, Inc., New Market, Maryland, was the principal author and project manager for this project. After his (much too premature) death, Albert Ness of the CCPS completed editing. Sandra A. Baker was co-author and editor. Brian Kelly, CCPS contributed case studies used in Chapter 2.

CCPS also gratefully acknowledges the comments submitted by the following peer reviewers:

- Salvador Avila - CCPS
- Pedro A. Bonilla – CCPS
- Jeffrey S. Caudill - Marathon Petroleum Company
- Erin P. Collins – Jensen Hughes
- Scott A. Haney - Marathon Petroleum Company LP
- John W. Herber – CCPS
- David E Herrick - Universidad de los Andes (Colombia)
- Casey Johnson – Covestro LLC
- John Kusowski - Engineering and Technical Associates, Inc.
- Peter N. Lodal – Eastman Chemical
- Michell L. LaFond - Dow Corning
- William Mosier - Syngenta
- Juliana Schmitz - Praxiar
- David Thaman - PPG

Their insights, comments, and suggestions helped ensure a balanced perspective for this book.

PREFACE

The American Institute of Chemical Engineers (AIChE) has been closely involved with process safety and loss control issues in the chemical and allied industries for more than four decades. Through its strong ties with process designers, constructors, operators, safety professionals, and members of academia, AIChE has enhanced communications and fostered continuous improvement of the industry's high safety standards. AIChE publications and symposia have become information resources for those devoted to process safety and environmental protection.

AIChE created the Center for Chemical Process Safety (CCPS) in 1985 after the chemical disasters in San Juanico, Mexico, and Bhopal, India. The CCPS is chartered to develop and disseminate technical information for use in the prevention of major chemical incidents. The center is supported by more than 180 chemical process industries (CPI) sponsors who provide the necessary funding and professional guidance to its technical committees. The major product of CCPS activities has been a series of guidelines and concept books to assist those implementing various elements of a process safety and risk management system. This concept book is part of that series.

The AIChE CCPS Technical Steering Committee recognized a significant increase in members' concerns about normalization of deviation over the past five years. During audits and inspections, findings showed issues related to normalized deviance. For example, some members found sporadic evidence that some process drawings and other process safety information (PSI) items were not always updated in a timely fashion. Operating procedures, maintenance procedures, safe work practices, and training modules were not always subjected to management of change (MOC) when the plant configuration changed in a way that affected their content. These specific examples of deviations from adhering to established work flow processes in management system administration and implementation may not result in immediate catastrophic release, but are serious concerns. When it comes to the physical day-to-day work practices of operators, maintenance technicians, and engineers, normalized deviance creeps in when skipping a step or changing a step's performance in a way that is not exactly as the approved work process describes becomes commonplace. The CCPS Technical Steering Committee initiated the creation of this concept book to assist facilities in recognizing and addressing the phenomenon of normalization of deviation.

EXECUTIVE SUMMARY

We have met the enemy and he is us.
Pogo (by Walt Kelly)

Normalization of deviance can affect any organization, for profit, non-profit, or governmental organizations, and manufacturing or service organizations. Even a highly reliability organization (HRO) can be affected by normalization of deviance.

There is a story about the Czar's courtyard. The Czar notices that there are two guards posted at each park bench in the royal courtyard. Why? Because 5 years before, the benches had been painted and the guards were posted to prevent the royal children from touching the wet paint. The order was never rescinded.

Leaders need to create a questioning culture, one where asking is not considered insubordination or blasphemy. Your behavior and actions as a leader are amplified to other employees as both an individual and as if they were the behavior and actions of the organization.

Normalized deviance, in Professor Diane Vaughan's words, is "(when) people within the organization become so much accustomed to a deviant behavior that they don't consider it as deviant, despite the fact that they far exceed their own rules..."

Employees at all levels acclimatize to the deviant behavior the more often they see it or do it.

- Outsiders see that the activities seem deviant.
- People within the organization see the deviance as normal behavior.
- In hindsight (that is, when caught), people within the organization realize that their seemingly normal behavior was deviant.

Pro-tip – Realize deviant behavior in foresight. Take appropriate action.

Other employees evaluate all decisions and associated actions by executive leadership, management, and supervisory employees, as representative of the organization's acceptance, or non-acceptance, of normalized deviance. It is perception based. People in leadership roles should:

- Take the time to evaluate each leadership decision against site, corporate, industry, and governmental guidance, rules and regulations.
- Make your decisions accordingly.
- Identify discrepancies in organizational systems and the organizational behavior that telegraphs the organization's acceptance level for normalized deviance.
- Reach consensus on the actions and prepare follow through actions to mitigate or support the anticipated response.

Actively recognizing and reducing normalization of deviation supports your organization's successful implementation of techniques such lean manufacturing, six sigma, total quality management, international organization of standardization, and others. Establish a management element to address recognition and reduction of normalized deviance within your business process systems for worker safety, process safety, environmental, and quality management.

1. INTRODUCTION

Climate is what we expect, weather is what we get.
Mark Twain

Individual and collective human behavior displays traits that are, in certain ways, analogous to the natural appearance and impact of weather events. Normalization of deviation is this type of human behavioral trait.

To help make the connection to this analogy, let us accept the following premise. If humans are involved in any type of goal reaching behavior, that activity involves a process. If a process exists, however formal or informal it may be, there can be deviation from that process. Whether we use a simple example, a person preparing their evening meal, or a complex example, a group designing, building, operating, and maintaining a chemical processing plant, the trait of normalization of deviation will undoubtedly appear in some measure. The consequences of such behavior need to be addressed.

- We always hope the weather (that is, our situation) will be fine today. Sometimes we have advance notice of what to expect from a forecast for the weather. This is similar to the chemical plant's most recent audit report with action items. Sometimes an unexpected weather event affects us (as will the effects of our or others' behavioral deviance).
- Normalization of deviance, like the weather, will appear and it will influence our experience. It is inevitable and can affect our goals in a negative way—unless we monitor it and react to it.
- The analogy works in the following way.
- The weather is sometimes beautiful all day and night. We wish it were like this every day. The process is running normally. Normalized deviance appears to be non-existent. Procedures and standards are being followed. All is well.
- On other days that start out beautifully, normalized deviance may appear later with low impact, like a nice day with a few clouds or showers that roll through the area. Our process is experiencing some upsets that need to be watched or addressed. We do what we need to do to avoid the immediate effects and things might clear up for the rest of the day.

- On days when the skies turn threatening, normalization of deviation's impact can be like a sudden storm, unanticipated and harmful. It can appear quickly and demand immediate actions not previously part of the planned process. It is like a violent thunderstorm or tornado that appears during a large outdoor public event. A process safety incident or near miss that results from non-compliance with standards would be a processing related example.
- On yet other beautiful days, maybe even on the very day described in the first bulleted example in this list, deviations are being normalized somewhere just outside of our awareness (more likely, right under our noses) and its effects are not sensed at all—or they are ignored like small changes in the weather—until specific events line up at the right time to allow a perfect storm to occur. Then, like tornados or tsunamis, process safety incidents can develop quickly.

Luckily for the chemical processing industry, the analogy above fails in one critical way. Unlike natural weather events, in the case of normalized deviance we can forestall the worst effects of the gradual organizational acceptance of nonconformance when we apply the twenty process safety elements of Risk Based Process Safety (RBPS) [CCPS 2007]:

Process Safety Culture: The combination of group values and behaviors that determine the manner in which process safety is managed. A sound process safety culture refers to attitudes and behaviors that support the goal of safer process operations.

Compliance with Standards: Identify, develop, acquire, evaluate, disseminate and provide access to applicable standards, codes, regulations, and laws that affect a facility and/or the process safety standards of care that apply to a facility.

Process Safety Competency: Maintain, improve, and broaden knowledge and expertise.

Workforce Involvement: A series of activities that (1) solicit input from the entire workforce (including contractors), (2) foster a consultative relationship between management and works at all levels of the organization, and (3) help sustain a strong process safety culture.

Stakeholder Outreach: The efforts to (1) seek out and engage stakeholders in a dialogue about process safety; (2) establish a relationship with community organizations, other companies and professional groups, and local, state, and federal authorities; and (3) provide accurate information about company/facility operations, products, plans, hazards, and risks.

Process Safety Knowledge: The work activities to gather, organize, maintain, and provide information to other process safety elements. Process safety knowledge primarily consists of written documents such as hazard information, process technology information, and equipment-specific information.

Hazard Identification and Risk Analysis (HIRA): All activities involved in identifying hazards and evaluating risk at facilities, throughout their life cycle, to make certain that risks to employees, the public, or the environment are consistently controlled within the organization's risk tolerance.

Operating Procedures: Written, step-by-step instructions and information necessary to operate equipment, compiled in one document including operating instructions, process descriptions, operating limits, chemical hazards, and safety equipment requirements.

Safe Work Practices (SWP) practice: An integrated set of policies, procedures, permits, and other systems that are designed to manage risks associated with non-routine activities such as performing hot work, opening process vessels or lines, or entering a confined space.

Asset Integrity and Reliability: A process safety management system for ensuring the integrity of assets throughout their life cycle.

Contractor Management: A system of controls to ensure that contracted services support (1) safe facility operations and (2) the company's process safety and personal safety performance goals. It includes the selection, acquisition, use, and monitoring of contracted services.

Training and Performance Assurance: Practical instruction in job and task requirements and methods, and the means by which workers demonstrate that they have understood the training and can apply it in practical situations. Training may be provided in a classroom or at the workplace, and its objective is to enable workers to meet some minimum initial performance standards, to maintain their proficiency, or to qualify them for promotion to a more demanding position.

Management of Change (MOC): A management system to identify, review, and approve all modifications to equipment, procedures, raw materials, and processing conditions, other than replacement in kind, prior to implementation to help ensure that changes to processes are properly analyzed (for example, for potential adverse impacts), documented, and communicated to employees affected.

Operational Readiness: The efforts to ensure that a process is ready for start-up/restart. This element applies to a variety of restart situations, ranging from restart after a brief maintenance outage to restart of a process that has been mothballed for several years.

Conduct of Operations: The embodiment of an organization's values and principles in management systems that are developed, implemented, and maintained to (1) structure operational tasks in a manner consistent with the organization's risk tolerance, (2) ensure that every task is performed deliberately and correctly, and (3) minimize variations in performance.

Emergency Management: The work activities performed to plan for and respond to emergencies. : A systematic approach for determining the causes of an incident and developing recommendations that address the causes to help prevent or mitigate future incidents.

Measurement and Metrics: Establishment of performance and efficiency indictors to monitor the near real-time effectiveness of the RBPS ,management system and its constituent elements and work activities.

Auditing: A systematic, independent review to verify conformance with the RBPS elements via a well-defined review process to ensure consistency and to allow the auditor to reach defensible conclusions.

Management Review and Continuous Improvement: The routine evaluation of whether management systems are performing as intended and producing the desired results as efficiently as possible.

By working to apply the concepts in this book thoroughly, you will find that occasionally a normalized deviation may reveal a positive thing. It occasionally reveals a process that needs to be revised to reflect reality. For example, many plants find that their operators are not using the operating procedures in the way desired by the company and expressed through rules. The reason is often that the documents are not current and the operators have been taking different actions to compensate. In addressing a discovery like this, it is essential that the organization evaluate these deviations on the part of the operators and assess how to best integrate them into various levels of the organization. These levels include management (technical and business), operations, maintenance, engineering, technical services, and quality. In this case, when an audit team finds the workers are not using the procedures, the team needs to dig deeper. The real deviation almost certainly stemmed from not managing change effectively. One good practice for this specific situation is to have provisions for the operators to recommend or make field changes to documents as they use them. When the field change process is planned and documented, it is a subroutine of your management of change (MOC) system. The need for a field change is identified when a worker actually uses a procedure to do a task, but finds the plant configuration different from that described in the procedure. It is an unquestionable opportunity for organizational learning to advance.

This book addresses how to recognize and respond to the normalization of deviation that can occur in any ongoing process that involves humans. The primary focus is on reducing the incidence of normalization of deviation and the associated increased risk exposure due to its effects while operating a chemical manufacturing facility. This book attempts to make it clear that actively addressing normalization of deviation can assist manufacturers to succeed in improving performance in five main business-driven areas:

- Process safety performance
- Personnel safety performance
- Environmental responsibility
- Product quality performance
- Sustainable long-term profitability

1.1 THE DEFINITION OF NORMALIZATION OF DEVIANCE

This book focuses on how normalized deviance relates to catastrophic events, but the techniques it offers to address normalization of deviation can optimize any business process—also called work processes—whether the focus of the process is personnel safety, quality, environmental compliance, or business viability.

The following definition from the *Guidelines for Risk Based Process Safety* provides the basis of our discussion on how normalization of deviation affects our industry, especially in the area of process safety management [CCPS 2007]:

Normalization of Deviation – A gradual erosion of standards of performance as a result of increased tolerance of nonconformance.

Normalization of deviation is a long-term phenomenon in which individuals, work teams, and entire organizations sometimes gradually accept a different standard of performance until that becomes the norm. It is typically the result of conditions slowly changing and eroding over time. A human interaction based deviation has to have occurred repeatedly, over time, without causing an incident or a problem that noticeably affects the process in other ways, such as yield or product quality. To summarize, normalization of deviation requires these characteristics (Figure 1.1):

- It is a human based deviation.
- The deviation occurs repeatedly, over time.
- The deviation does not cause an immediate incident or a noticed process effect.

This does not necessarily mean the same work process deviation is occurring repeatedly, but the same types of deviations are occurring repeatedly; skipped steps, workarounds, and shortcuts. Nor does a deviation have to be practiced consistently. These types of deviations may be commonly practiced in the same way by one work crew "Hey, I found a new way to change that filter cartridge! ", but they may not be communicated between work crews. Some people will do it one way, some will do it another way.

Normalization of deviance can begin as a shortcut or poorly documented temporary change in a standard work practice, procedure, or business process. If there are no apparent negative consequences, or there is no recognition of the change as a deviation from the standard, the new practice becomes accepted and displaces the original practice. Over time, this process repeats. When the changes are small and seemingly insignificant, they are easy to miss. Normalization of deviance in critical process steps for hazardous chemical units may not always represent a breach of the management of change (MOC) system. It includes shortcuts related to management of subtle changes, even when an item may be considered a replacement-in-kind.

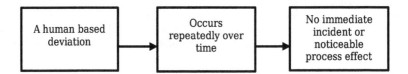

Figure 1.1. Three Characteristics of Normalized Deviation

The start can be a deviance within the acceptable operating range just on the low or high end and over time drift out of the acceptable range and become a safety issue. By that time the normalization of deviance is already established.

These types of deviations have been major contributors to many serious incidents. As an example, the Baker Panel Report found normalized deviance contributed to organizational dysfunction within BP [Baker 2007]. One example included a 2005 behavioral safety culture assessment that a consultant conducted described "the occurrence of at-risk behaviors and normalized deviations" at the BP Toledo refinery. The consultant further described a perception among workers that safety was not the highest priority among the refinery workforce. The consultant noted a belief that equipment was run to failure or near failure and problems were not fixed "until something bad happens."

The BP Carson refinery appeared to have some issues related to operating discipline, risk identification, toleration of deviations from safe operating procedures, and apparent complacency toward serious process risks. A third-party behavioral safety culture assessment that BP commissioned indicated that the Carson refinery suffered to some extent from normalized deviations from safety practices, although the assessment found these deviations to be more common in personal safety procedures. It also found a number of factors driving at-risk behavior at Carson:

1. a culture that did not accept failure
2. the normalization of deviations from safety practices through repetition and the resulting failure to either recognize or report risk
3. positive reinforcement of unsafe behaviors as well as the lack of positive reinforcement of safe behaviors
4. the open-endedness of procedures that invited individual interpretation

1.1.1 Deviation Versus Deviance

Is there a difference between normalization of deviation and normalized deviance? The terms are used interchangeably in general discussion of the concept and in this book. Explicitly, deviation means departing from an established course or accepted standard (and in statistics, it means the amount by which a single measurement differs from a fixed value such as the mean). Deviation implies a somewhat more quantitative departure (think standard deviation) than deviance does. It implies that something can be isolated and counted or individually examined. Usage examples are, *"A contributing factor to the incident was a deviation from the normal pressure range"*, or *"Three deviations from the approved procedure were observed."*

Deviance, however, implies a more general quality. The definition is the state of departing from usual or normal behavior. A usage example would be, *"Our audit team found deviance in the management of change / pre-startup safety review elements of the administrative program."* Deviance refers to the fact or state of departing from usual or accepted standards. Deviance can also be used to denote a systemic compounding of multiple or repeated deviations.

Regardless of which term or combination of terms is used in general discussion or throughout this book, the point remains that individual deviations to standardized work processes that go unrecognized or uncorrected will lead to deviance in those same work processes until eventually the deviations replace the standardized work processes.

1.2 THE MOTIVATION FOR WRITING THIS BOOK

While the CCPS team found several overarching normalization of deviation issues in the chemical processing industry that motivated this book's development, the ubiquity of normalized deviance within all organizations is, in itself, reason to address it. Other inherent issues include:

- Normalization of deviation can have a dramatic impact on production efficiency, product quality, process cycle times and other economic concerns.
- The unexpected appearance of normalized deviation promotes human error during operational and incident response efforts.
- The effects of normalized deviation can range from a quality issue to a catastrophic event.
- Incident investigations regularly identify both catastrophic and near miss process safety events where normalization of deviation was a contributing factor.
- Recognizing normalized deviance can be a significant improvement tool to help the chemical industry monitor every element of a process safety management system.
- Economic catastrophe after a process safety incident is a real possibility for some sectors of the chemical processing industry. Normalization of deviation can cause catastrophes.

1.3 OUR AUDIENCE AND HOW TO USE THIS BOOK

This book is intended for anyone interested in normalization of deviance as it relates to the chemical processing industry specifically or to manufacturing in general whenever there are critical process safety, personnel safety, quality, or environmental implications. The Center for Chemical Process Safety (CCPS) seeks to help both experienced process safety professionals and persons entering the field of process safety to better understand and manage normalization of deviation.

This book offers examples of techniques used by industry leaders to identify, respond to, and alleviate normalization of deviation. It can be used to check your practices and help reinforce or improve upon your current process safety management system.

The following list identifies positions within an organization that may benefit from this book:

- Managers of process safety and risk management programs and process safety management (PSM) coordinators at a manufacturing facility
- Corporate process safety management staff
- Project managers and project team members whose projects initiate the PSM life cycle for their design
- Engineers or other staff members performing management of change activities, particularly final approvers
- Incident investigation team leaders, team members, and trainers
- Industrial hygienists and safety department personnel
- Operations, maintenance, and other manufacturing personnel who may be part of a PSM element team
- Any employee participating in the PSM program
- Contractors at chemical processing facilities
- Executives and business leaders of any organization in the chemical process industry (CPI) and outside the CPI. Understanding the concept of normalized deviance is helpful in all business processes

1.4 HOW OUR WORLDVIEW AFFECTS US WHEN RECOGNIZING NORMALIZED DEVIANCE

Diane Vaughan is an American sociologist with wide-ranging academic interests. She developed the concept of normalization of deviance to describe a phenomenon she found through her work looking at the darker side of organizations. Her primary interest is in how things go wrong. *How did we make that huge mistake? How could that person have done such an unethical/illegal/incorrect thing? How did we let this disaster happen?* Her research points out that trouble comes not only from individual performance failures but also from endemic organizational failures.

Her work gained worldwide attention from the process safety community, incident investigators, and many other readers when she published *The Challenger Launch Decision* in 1996 [Vaughan 1996]. After the Columbia space shuttle lost mechanical integrity upon reentry in 2003, Professor Vaughan was invited to join the Columbia Accident Investigation Board (CAIB). Her participation on the team helped reveal that the NASA organization still had not learned from the 17-year-old, and deeply investigated, mistakes of the Challenger disaster. Unacceptable risks had been taken. Normalized deviance had crept back into NASA's daily practices regarding hazardous operations in spite of an ultimate lesson-learned experience.

During an interview, Dr. Vaughan described the concept of "social normalization of deviance." This is when people are so used to deviance that they don't consider it as deviance, even if the deviant behavior violates the organization's own rules for safety. The problem with social normalization is that workers get used to it and do not even think they are deviating from the rules or procedures [Vaughan 2008].

1.4.1 Regulatory Influences

Regulations and industry guidelines can influence our worldviews regarding deviance from them. In *Controlling Unlawful Organizational Behavior: Social Structure and Corporate Misconduct,* Dr. Vaughan notes the nature of the rules and laws we establish may influence deviations from those rules [Vaughan 1983].

Her examples of how regulations and rules affect deviation follow:

1. Number: Ignoring a large number of guidelines related to an industry may defy mastery or result in some regulations. Large numbers of laws and rules are difficult to monitor which reduces the likelihood of detecting deviations.
2. Recency: The age of a regulation or law may affect its legitimacy, knowledge of its existence, or the number of interpretations and precedents available for guidance. These factors can affect the desire to conform or deviate.
3. Relevance: Whether the organizational consensus agrees with the law in comparison with the larger purpose of their goals can influence willingness to comply.
4. Complexity: When regulations and laws are very complex and interrelated, they can be subject to different interpretations. It can cause deviance through simple misunderstanding.
5. Vagueness: Guidelines or laws that are unspecific or unclearly written can cause deviation from the rules' intent.
6. Acceptability: Does the facility or industry accept the rule? Does it match their focus; is it costly to adhere to in terms of resources? The

strength and consistency of enforcement fines and sanctions can affect deviation from the regulations.

Consider how the chemical industry views the nature of process safety rules around the world in light of the six categories above. What is the likelihood, or risk factor, that your facility or company is pushed toward deviation due to the six dimensions listed above?

1.4.2 Unique Worldviews

Dr. Vaughan's statement on the "Social normalization of deviance..." is about the individual and organizational worldviews of the people within the organization. That would be the chemical and petrochemical process industry. We are constantly evaluating our perceptions of the universe to create our individual conception of the world. As it is based upon perception, it varies from person to person and organization to organization as well as in every team or department. Additionally, it refers to the framework of ideas and beliefs forming a global description through which an individual, group, or culture watches and interprets the world and interacts with it. Other entities—humans and organizations—seem to be the most interesting subjects for us to evaluate. A special personality state is needed for us to know ourselves well enough to critique our personal mental models and identify when they need alteration.

The same case applies to organizations. The internal politics and status issues of the organization, which is made up of individuals who work in the boardroom as well as those who work in the boiler room, make it difficult for any organization to hold cohesive, organizational self-awareness. There is one holistic worldview held by the group, but it is often difficult to describe easily from inside the group. The group culture is best determined through observing displays of it in group decisions and actions.

Cultural influences can also affect people's worldview and normalization of deviance. Some influences include:

- **World and regional influences.** Particularly important for a large multinational, integrated company.
- **Local influences, within one country or continent.** Even within a country, depending on where you are in the country, the regional environment can influence personal risk tolerance. For example, in the United States, not all states require compliance with National Fire Protection Association (NFPA), or the National Board Inspection Code (NBIC) for boilers and pressure vessels. Design, maintenance, and inspection practices can vary significantly between states that require code compliance and those that do not. This can become a catalyst for acceptance of deviance within the workplace.
- **Rural versus Urban influence.** Plant management for plants located in rural areas (particularly where the workforce comes from an agricultural background) report that they need to essentially reprogram new hires' mindsets. Many new workers have spent their lives being rewarded for finding shortcuts. In an agriculture based economy, shortcuts equal more

food on the table, or money in the bank. The desire to take uncontrolled shortcuts must be prohibited in the workplace.

- **Economic influences.** Socioeconomic issues come into play at both the personal and organizational level. The capital or operating budget may be squeezed by some other force, such as the market or an internal drive for increased profit. Also, are the workers receiving equitable pay for the effort and skills they bring?

How do we determine when we are experiencing a normalized deviation at our facility? Is our individual worldview in the way? Is our organizational worldview skewed because of a long run of good luck or a brand that implies our organization is excellent and can do no wrong? Is it such a small change from our process that we justify it in our own minds and clear a path for normalized deviance to progress?

1.5 WORK PROCESS KNOWLEDGE IS ESSENTIAL IN DETERMINING THE EXISTENCE OF DEVIATION

If you can't describe what you are doing as a process,
you don't know what you're doing.

W. Edwards Deming

This quote could be considered a description of the foundation of process safety as a concept. Deming's words emphasize the proposition that an organization must know all of its business processes, from manufacturing operations to back office, in order to truly understand them and then try to improve them. For example, you must have process knowledge of all types in order to build or improve a process safety management system; process safety information, business process information, and accurate real-time process data are examples. Easy access to accurate process information supports process safety improvements, environmental performance, product quality, and provides for long-term economic viability.

The optimal time to identify and address normalization of deviance is the first time a work process deviation that had no immediate consequences is noticed but, by our definition and means test, normalized deviation as such has not truly occurred in the single first instance. The first deviance that is noticed and ignored is the first step toward a normalized deviation. In order to recognize deviations at your facility, you must know the requirements of the processes that make up your total business management process. From a process safety standpoint, we can then focus upon those managerial, supervisory, administrative, safety, environmental protection, operating, maintenance, shipping and related processes critical to preventing the release of highly hazardous chemicals. Chapters 4, 5 and 6 will describe techniques to recognize and deal with deviance. To do this you first require a well-designed integrated process safety management system for a facility.

1.6 NORMALIZED DEVIATION AND TRADITIONAL PROCESS SAFETY CONCEPTS

How does normalization of deviation tie into traditional process safety concepts? Every process, whether an accounting process, a sales process, a management system process, a chemical plant unit operation, or a maintenance process, presents opportunity for deviation from that process.

A simple incident causation model that applies to normalization of deviation is the Hazard-Barrier-Target (HBT) theory (also called the Swiss Cheese Model) [Reason 1990]. This model provides an interesting view of the multiple-cause or system theory of incident causation. In HBT, an investigator starts with the understanding that a process has one or more inherent hazards. The hazard is a property of the process such as toxicity of a chemical, stored energy such as pressure much higher or lower than ambient, and electrical hazards. The target can be a person or the environment, and in an abstract sense, some interpret the target to be any loss impact. For example, the target could be product and lower quality could be the impact. The barriers are actually layers of protection that prevent the hazard from having a negative impact on the target. One important concept stressed in HBT is that ALL barriers have weaknesses; therefore, each barrier has some probability of not working when needed. These weakness are represented as a hole in the barrier. The most important concept for any incident investigator to learn may be the following statement:

No layer of protection is perfect.

Incidents occur when all barriers fail to prevent harm and a near miss occurs when one or more barriers fail. HBT is an excellent teaching tool for incident mechanisms and for describing the probabilistic nature of incidents, even for protected systems. For the purposes of this book, a deviation that could escalate to normalized deviance is the removal or neutralization of at least one barrier so that the hazard is more likely to affect the target [CCPS 2003].

Traditional process safety concepts that drive process safety excellence include the following [Amyotte 2011]:

- Risk management
- Inherently safer design
- Human error and human factors
- Safety management systems
- Safety culture

Normalized deviance can occur when implementing any of these general concepts. However, when programs are well designed, effectively implemented, and audited for compliance, they can:

- work to reduce the frequency of normalized deviations
- find deviations from the normal processes
- help abate normalized deviances found
- let us know when our normal process needs to be changed

1.6.1 Process Safety Around The Globe

Worldwide recognition of the benefits of process safety to workers and the environment became obvious beginning in the 1980s. Many governments and industry groups adopted process safety related standards or guidelines, continuing the sharing of information. Table 1.1 lists organizations and an example document they produced to recommend or require process safety elements be implemented.

For the purposes of this book and our discussions of deviance from administrative processes, we will use the model of the twenty CCPS RBPS elements [CCPS 2007]. In addition, there are specific regulations and directives that may apply to different locations (some examples are noted in Table 1.1). To keep this book's scope manageable, this book will note the similar regulatory elements in the *U.S. OSHA Process Safety Management* (PSM) standard and in sections of the *U.S. EPA Risk Management Program (RMP) Compliance Plan*. The following list summarizes the core of these U.S. PSM/RMP regulations [OSHA 1992, EPA 2017]:

- **Develop and maintain written safety information** identifying workplace chemical and process hazards, equipment used in the processes, and technology used in the processes.

- **Perform a workplace hazard assessment**, including, as appropriate, identification of potential sources of releases, identification of any previous release within the facility that had a potential for catastrophic consequences in the workplace, estimation of the effects of a range of releases, and estimation of the health and safety consequences of such a range on employees.

- **Consult with employees and their representatives** on the conduct of hazard assessments, and incident prevention/process safety plans. Provide access to these and other records required under the standard.

- **Establish a system to respond to the workplace hazard assessment findings**, which need to address prevention, mitigation, and emergency responses.

- **Periodically review the workplace hazard assessment** and response system.

- **Develop and implement written operating procedures for the chemical processes**, including procedures for each operating phase, operating limitations, and safety and health considerations.

Table 1.1 Examples of Global Process Safety Regulations and Guidance

Note. Please consult your local region or jurisdiction for specific requirements.

Australia Australian National Standard for Control of Major Hazard Facilities	Australian National Standard for the Control of Major Hazard Facilities, NOHSC: 1014, 2002. www.docep.wa.gov.au/
Canada Canadian Environmental Protection Agency, Environmental Emergency Planning	Environmental Emergency Regulations (SOR / 2003-307), Section 200, Environment Canada. www.ec.gc.ca/CEPARegistry/regulations
China Guidelines for Process Safety Management	Guidelines for Process Safety Management, AQ/T3034-201o; Effective 01-May-2011.
Europe European Commission Seveso III Directive	Control of Major-Accident Hazards Involving Dangerous Substances, European Directive Seveso-III (Directive 2012/18/EU). ec.europa.eu/environment/seveso/legislation.htm
France Ministry of Interior Orsec	Orsec (Organisation de la réponse de sécurité civile). Translated: Organization of the civil protection response. www.interieur.gouv.fr/Actualites/Dossiers/Le-plan-Orsec-a-60-ans.
Korea Korean Occupational Safety and Health Agency, Process Safety Management	Korean Occupational Safety and Health Agency, Industrial Safety and Health Act, Article 20, Preparation of Safety and Health Management Regulations. Korean Ministry of Environment, Framework Plan on Hazards Chemicals Management, 2001-2005. english.kosha.or.kr/main
Malaysia Department of Occupation Safety and Health Ministry of Human Resources Malaysia	Malaysia, Department of Occupational Safety and Health (DOSH) Ministry of Human Resources Malaysia, Section 16 of Act 514. www.dosh.gov.my/doshV2/
Mexico, Secretary of Labor and Social Welfare	Secretary of Labor and Social Welfare = Secretaría del Trabajo y Previsión Social (STPS), Mexican Standard NOM-028-STPS-2012, System For The Occupational Administration – "Safety In The Critical Processes And Equipment That Handle Hazardous Chemical Substances."
United Kingdom Health and Safety Executive COMAH Regulations	Control of Major Accident Hazards Regulations (COMAH), United Kingdom Health & Safety Executive (HSE), 1999 and 2005. www.hse.gov.uk/comah/
United States EPA Risk Management Program (RMP) Regulation	Accidental Release Prevention Requirements: Risk Management Programs Under Clean Air Act Section 112(r)(7), 40 CFR Part 68, U.S. Environmental Protection Agency, June 20, 1996 Fed. Reg. Vol. 61[31667-31730]. www.epa.gov
United States OSHA Process Safety Management (PSM) Standard	Process Safety Management of Highly Hazardous Chemicals (29 CFR 1910.119), U.S. Occupational Safety and Health Administration, May 1992. www.osha.gov

- **Provide written safety and operating information for employees and employee training** in operating procedures, by emphasizing hazards and safe practices that must be developed and made available.

- **Ensure contractors and contract employees** are provided with appropriate information and training.

- **Train and educate employees and contractors in emergency response procedures** in a comprehensive manner.

- **Establish a quality assurance program** to ensure that process-related equipment, maintenance materials, and spare parts are fabricated and installed consistent with the initial design specifications. (See RBPS Asset Integrity Management)

- **Establish maintenance systems for critical process-related equipment**, including written procedures, employee training, appropriate inspections, and testing of such equipment to ensure ongoing mechanical integrity. (Note: RBPS Asset Integrity element)

- **Conduct pre-startup safety reviews** of all newly installed or modified equipment. (Note: RBPS Operational Readiness element)

- **Establish and implement written procedures for managing change** to process chemicals, technology, equipment, and facilities.

- **Investigate every incident that results in or could have resulted in a major release** in the workplace, with any findings to be reviewed by operating personnel and modifications made, if appropriate.

- **Develop an emergency response plan for protecting the public and the environment** and coordinate activities with the community emergency planners. Facilities whose employees are expected to respond to releases of hazardous substances must develop an emergency response plan for protecting the public and the environment and must coordinate their activities with the community emergency planners and responders. Facilities whose employees will not respond do not have to prepare an emergency response plan; however, they must have an appropriate mechanism in place for notifying emergency responders in case of an incident.

It is important to note that normalized deviance can occur in any organization anywhere in the world, no matter which regulation, directive, or corporate standard applies.

Management system documents are written by an organization with the intent of establishing minimum standards and recommended internal guidelines to support performance based compliance with the applicable regulations. These documents become organizationally imposed rules. Many of them address very prescriptive aspects of the regulation, but also must address the choices an organization has in how it implements the process safety management system. When designing the system one should address the organizational culture and establish the resource levels that will be applied. Work processes are more likely to be followed when they match the organizational culture and make sense to the person implementing them.

When this group of interwoven documents is developed with the cooperation and consensus of the workers, technical staff, and a representative of every department or group that has a role in implementing a process safety element, it provides a tapestry of enhanced risk management. The audit team needs to pay special attention when they recognize that a deviation may actively be normalized within the organization in regard to this program.

A well-documented and maintained process safety management system will augment organizational memory. Up-to-date piping and instrument diagrams (P&IDs), accurate procedures, hazardous area classification drawings, and a thoroughly documented and implemented management of change and pre-startup safety review system are prime examples of how the system creates memories. Maintenance of organizational memory should be reinforced continuously throughout the life cycle of a processing facility.

Maintaining accurate configuration data in real time and holding regular emergency drills are common practice in the commercial nuclear power industry. This promotes the concept of a healthy sense of vulnerability. Management encourages everyone to maintain a calm sense of awareness that the worst-case scenario can happen right now and that we need to be ready to follow our response procedures. Extensive training, strict document control for all critical procedure performance, configuration management, and other good practices keep this in the front of each employee's mind. A healthy sense of vulnerability is an attitude you should encourage and develop within your organization. Do not rely on a history of no catastrophic incidents as a sign of good performance. That history may just be good luck.

Table 1.2 gives some examples of how the RBPS PSM elements fight normalization of deviation element by element. You should be aware, however, that a PSM system could become a cause of deviance if it becomes too cumbersome. Solutions to this systems issue are described in Appendix G, *Guidelines for Management of Change for Process Safety* [CCPS 2008a].

When these specific concepts are implemented effectively as an interwoven system, they work to reduce the frequency of normalized deviation occurrences, find deviations from the normal processes, help abate the normalized deviances found, and occasionally, let us know when our normal process needs to be changed.

A movement over the past few decades toward developing an integrated management system approach in the refining and chemical processing industry (sometimes referred to as an operational excellence management system) has shown how these traditional process safety methods serve to support all of the other drivers and their needs. The overlap provides leverage for building a total business management system. *Guidelines for Integrating Management Systems and Metrics to Improve Process Safety Performance* covers integrated management systems in more detail [CCPS 2016a].

Table 1.2 Process Safety Elements: How They Reduce Normalized Deviance

Process Safety Element	How it Reduces Normalized Deviance
Process Safety Culture	A good process safety culture enables an organization to establish a high standard of performance and enables effective implementation of the other PSM elements. Employees will know the organization values process safety and will provide the resources to control risks.
Compliance with Standards	Compliance with standards will help a company design, operate and maintain a safe facility by taking advantage of the knowledge contained in the codes and complying with regulations.
Process Safety Competency	Competent people can transform information into knowledge. This allows people at all levels in the organization to understand and manage the process risks.
Workforce Involvement	Involving employees at all levels in new process safety program development or in maintaining and overseeing mature plans raises overall awareness of the level of operational discipline needed for rigorous process safety. Participants understand the "whys" of the administrative program better. They are more likely to communicate error up through the organization when it is perceived as a valued behavior.
Stakeholder Outreach	This element includes establishing relationships with professional groups such as the CCPS, American Petroleum Institute (API), International Electrotechnical Commission (IEC), IChemE, European Process Safety Centre (EPSC), etc. and encourages sharing of information and lessons learned with similar facilities and organizations.
Process Knowledge Management	Maintaining accurate process safety information (PSI) is a core information need for fighting normalized deviance. Not maintaining PSI effectively is often an example of normalized deviance. When allowed to continue, it can trigger separate deviations by inaccurate PSI and well-intended shortcuts taken by employees to address the PSI deficiency. Accurate PSI can discourage deviations like these.

Table 1.2 Process Safety Elements: How They Reduce Normalized Deviance (continued)

Process Safety Element	How it Reduces Normalized Deviance
Hazard Identification and Risk Analysis (HIRA)	Performing and maintaining accurate and up-to-date process hazard analyses provides core information needed for fighting normalized deviance. PHA helps identify the most critical physical and operational aspects of a process that normalized deviance can affect. It allows the organization's vigilance to be more focused on areas that represent the highest risk.
Operating Procedures	Writing and revising effective procedures gives workers reliable information to prevent deviations and their subsequent normalization. At a minimum focus on: operations procedures, safe work practices, maintenance procedures, and emergency response plan procedures. Other categories can be controlled as required by regulations. This also supports basic lean manufacturing techniques and consistent levels of operational excellence.
Safe Work Permits	This safe work practice, which applies in any manufacturing facility, can prevent very specific deviations related to common work practices involving control of ignition source, lock-out/tag-out, entry into confined spaces, etc.
Asset Integrity and Reliability	High quality maintenance of a well-designed physical process prevents process operation deviations. Workers will deal less with mechanically based operating issues that may encourage deviation. Also having a set of critical maintenance procedures effects a protection layer on the methods used to maintain equipment consistently. Training maintenance workers on the process and its hazards helps them have a better understanding of their roles in process safety.
Contractor Management	An effective program to select qualified contractor suppliers with good safety and work quality records, and training the contract workers on the unique hazards of the client facility can prevent imported deviations as well as newly developed homegrown ones. Imported deviations can come from a contractor's previous experience in facilities that may have had a very different risk profile.

Table 1.2 Process Safety Elements: How They Reduce Normalized Deviance (continued)

Process Safety Element	How it Reduces Normalized Deviance
Training and Performance Assurance Training Program	Providing effective training for new employees and updating existing employees with training on process changes that affect their jobs is a preventative measure against normalization of deviation. A formal training program minimizes the passing on of deviations from trainer to trainee. A well-trained employee will tend to confirm their knowledge when they encounter process related steps or conditions of which they are unsure. Encourage all employees to learn on their own and share what they find with the organization to improve each other's process knowledge and skills.
Management of Change (MOC)	Working with PSSR, the management of change element reduces the opportunity of introducing new deviations through thorough analysis, approval to proceed, and preparation for controlling permanent and temporary changes in process chemicals, technology, equipment, procedures, and process facilities. MOC provides for accurate data update and training and documented steps to help avoid the first deviation when starting up a change. Personnel understand the need to review a change before implementing it is a key part of catching deviations before they become part of normal practice.
Conduct of Operations	Conduct of operations includes, but is broader in scope than, operating discipline. It involves first planning and documenting the work to be done and then executing according to the plan. ("Plan the Work--Work the Plan"). Conduct of operations implies procedures are understood and followed, equipment is maintained, and changes are controlled. It also applies to engineering as well as operations. Design standards are followed, with up to date PSI and P&IDs are provided when the plant is handed over.

Table 1.2 Process Safety Elements: How They Reduce Normalized Deviance (continued)

Process Safety Element	How it Reduces Normalized Deviance
Operational Readiness	This element works with management of change to provide a two-tier check of any configuration changes from the current design. The purpose of the operational readiness review (commonly known as the Pre-Startup Safety Review (PSSR)) is to ensure that the change was implemented as approved in the MOC and that all of the required procedures and training have been accomplished so that the affected personnel are well prepared to begin operating and maintaining the changed process. It also documents that the organization is reducing the likelihood of introducing a deviation through analysis, study, and preparation for a change. Clear definition of PSSR critical deliverables (both new facilities and modifications to an existing facility) is the first step. A robust PSSR program is a hard stop. Evidence of completeness for each deliverable needs to exist prior to authorization to proceed to startup. Including an additional authorizer, one independent of the project, reduces opportunities to deviate.
Emergency Management	This element plans for reasonable contingencies in order to avoid deviation during a very stressful time for employees and the organization. Drills and training are essential pieces of this element as those present the only opportunity to notice deviations in practice prior to a live emergency.
Incident Investigation	In spite of effective performance in other elements, incidents can still occur at a site. Learning from incidents and near misses helps prevent repeat deviations that could become normalized. Thorough investigation also captures data on when normalized deviance was documented as a definite contributing factor to an incident. Near misses are often an early warning that a deviation has become normalized and a part of an organization's culture.

Table 1.2 Process Safety Elements: How They Reduce Normalized Deviance (continued)

Measurement and Metrics	You don't improve what you don't measure. Uses of metrics enable an organization to identify areas that need improvement, and can identify where normalization of deviance has occurred, for example, the organization tolerates late inspections or inadequate follow-up of recommendations from PHAs
Auditing	This element forces a period of organizational self-reflection regarding process safety performance. It focuses attention on the administration of the program itself and allows a team to look for and find examples of normalized deviations.
Management Review and Continuous Improvement	Routine management review of metrics, audit findings, incident and near miss investigations enables management to identify areas where normalization of deviance may be lurking.

1.6.2 Implementing a New PSM System

The book *Guidelines for Implementing Process Safety Management Systems* lists the following steps for creating a new PSM system [CCPS 2016b]:

1. Develop the design specification for the PSM system (by reviewing existing frameworks to determine the one preferred going forward). This includes performing a gap assessment between the PSM system to be implemented and existing processes and procedures.
2. Create element and system workflows (as appropriate).
3. Estimate the workloads and resources needed to implement the new elements and system.
4. Develop written programs and procedures for the elements and system.
5. Roll out the elements and system at a single site to act as a pilot program before rolling them out to the entire company.
6. Monitor implementation and initial performance, and modify the elements and/or system to make them work for the pilot site. Once the PSM system is rolled out to the entire company, monitor its progress every six months and share the results with management.

You are encouraged to use this book if you are creating or modifying a system within your process safety and risk management program. For example, Figure 1.2 illustrate an approach for developing and maintaining a Process Hazards Analysis (PHA) system. This approach applies to process safety management in the same way it works with quality management systems, environmental management systems, business management systems, and occupational safety management systems. This model can aid in creating an organization that can achieve sustainable development and long-term operability. Equally important, a well-designed PSM system can help an organization develop a culture that supports safety excellence.

Many corporations integrate their management systems for Process Safety, Personal Safety, Health, Environmental, Security, and Quality to streamline procedures. You are encouraged to read *Guidelines for Integrating Management Systems and Metrics to Improve Process Safety Performance* for a general approach to process safety program development and maintenance [CCPS 2016a].

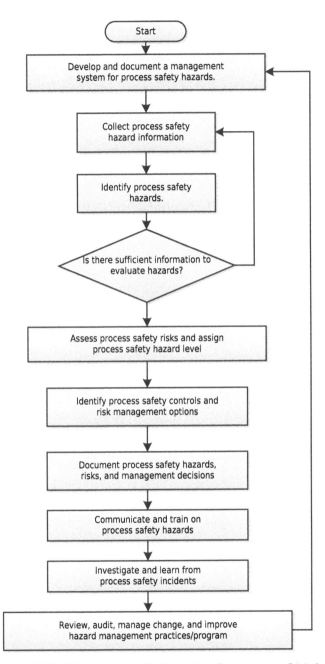

Figure 1.2. PHA Management System Development and Maintenance

2. WHY EXAMINE THE PHENOMENON OF NORMALIZATION OF DEVIATION?

*The single biggest problem in communication is
the illusion that it has taken place.*
George Bernard Shaw

2.1 INTRODUCTION

Bad news tends not to travel up the chain-of-command. Ultimately, this means that decisions on risk tolerance might not be made at the optimum management level. When this problem occurs, normalization of deviation has an open field in which to propagate. It can spread in two directions—width and depth.

- It widens across a work group when one deviation encourages its expansion and repetition by employees at every opportunity to deviate.
- It deepens across an organization as deviations are accepted by departments up and down the organization, such as manufacturing, maintenance, or engineering.

Employees repeat a deviation, altering it over time, until it has the potential to have a singular impact on an operation. The result is that the deviation reaches its maximum impact capability sooner, no matter whether the hazards and risks are high or low.

The presence and frequency of single deviations occurring across more processes and the probability of repetition of any of those deviations toward the point of normalization are both are likely to escalate. This leads to more negative outcomes. Three mindsets increase probability of deviation. The mindsets are:

- Unknowing of the hazard,
- Unbelieving of the probability, and
- Underestimating the severity of the outcome.

Unknowing of the hazard is the mindset when an employee (or organization) is not aware of the hazards of the process and the risk associated with process actions. This can be the result of not doing a hazard review, or a poorly done review. This can also be a case of a cognitive disconnect in a person who was made aware once, but simply became accustomed to the hazard over time. For example, a pressure safety valve (PSV) has a rupture disk underneath it to prevent fouling of the interior components of the PSV. A tell-tale pressure gauge is installed to indicate breach of the rupture disk. During a field review, it is noted that a tell-tale pressure gauge is indicating pressure. When the person in charge of the PSV program was asked when the rupture disk would be repaired the answer was, *"That thing is always failing; we will get to it at the next scheduled turnaround in six months."* The person did not understand that under certain circumstances a breached rupture disk could prevent the PSV from operating as designed. In addition, the regular failure of the rupture disk was not being addressed through design modifications or process operational changes. Deviations had been normalized.

Unbelieving of the probability is holding the belief that something cannot or will not happen. Sometimes this thought is held by those who understand technically how the event could unfold, but unconsciously block that knowledge from consideration. For example, a fire protection system risk review identifies that the underground fire main pressure is being maintained by the jockey pump at 50 psi instead of the designed 130 psi. The rationale given for operating at a reduced jockey pump pressure was because the fire main is prone to ruptures. No thought had been given to the fact that in a real fire situation the fire pumps would start automatically, increase the fire main pressure substantially, potentially rupturing and disabling the fire main. No funding to replace weaker sections of the fire main is sought. In other cases, the person may recognize that an event could happen but does not believe it will happen to them because of past good luck — *it has never happened here before.*

Underestimating the severity of the outcome is expecting a consequence that is less than the scenario's actual capability. This mindset occurs either through individual or organizational apathy or through human distraction due to outside influence. Outside influences can range from personal life issues to site culture, corporate policy, or immediate production needs.

For example, a third party warehouse is used to store oxidizers. After several audits, it becomes apparent that the warehouse manager is more interested in maximizing the warehouse capacity rather than managing the risks.

- Oxidizers are being stored in areas where sprinkler protection density is inadequate
- Incompatible materials are repeatedly stored amongst the oxidizers. The warehouse manager had been written up on this on previous audits.
- Storage practices are ignoring fire code limitations such as storage heights, pile size, aisle width, and distance from walls.

Potentially catastrophic deviations at this outsourced warehouse were normalized in spite of well-known hazards even though those hazards and their associated risk management had been addressed in communication with the contracting company prior to awarding the contract. This example may be a case of the *availability heuristic*. An availability heuristic is a mental shortcut that people build from their most immediate or recent examples that quickly come to mind when making a decision or evaluating a situation. The availability heuristic is simply stated as, "if a person can recall a thing, it has inherent importance." This term also implies that humans tend to weight their judgments toward those affected by more recent information, thereby making new opinions biased toward that latest news. The warehouse management team may have been making decisions based upon solving a recent problem that emphasized increasing shareholder value, i.e., stock price or dividends, over process safety (or other indicators of company performance) [Phung 2018].

The availability of consequences associated with an action is positively related to perceptions of the magnitude of the consequences of that action. In other words, the easier it is to recall the consequences of something the greater those consequences are often perceived to be. People often rely on the content they recall most easily unless they are challenged to bring more relevant (but harder to retrieve) material to mind.

2.2 PAST INCIDENTS RELATED TO NORMALIZED DEVIANCE

It is likely that every process safety incident in the history of mankind has a contributing factor that could be categorized as a human error or inadequate training. But, did the investigation find the normalized deviation?

When the analysis points to human error, companies can take one of several different approaches. In some cases, the human error or training deficiency is identified and evaluated. Many times, the management systems that contributed to the human error are not as thoroughly identified and addressed. Consider this example of a storage tank of flammable liquid being filled by an operator using a pump:

- The only safeguard against overfill is a sight-glass on the side of the vessel.
- The operator overfills the storage tank leading to release of flammable liquid to the flare via a PSV on top of the storage tank.
- As liquid flow into the tank continues, a second PSV that discharges directly to atmosphere opens resulting in loss of containment.
- An investigation team initially documents the root cause as human error.
- The incident report is not approved and a new team is sent to investigate.

- The new team identifies the process hazard analysis (PHA) as being inadequate.
- The PHA team had assumed an overflow situation would only discharge to the flare and no alarms or interlocks had been recommended based on this assumption.

It would have been wrong to reprimand the operator. The second investigation team identified that the PHA team had not considered continued flow into the tank would result in opening of the second PSV due to pump discharge pressure.

In other cases, human performance is translated back to the following of operating procedures or emergency response procedures and whether the procedures have the proper level of detail. The training that the employee received is also reviewed.

Factors such as fatigue and the influence of the human machine interface (HMI) layout on human performance are not typically identified as root causes. For example, the operator of a steam turbine generator has one computer screen in front of him and another older computer screen behind him. During a process upset and emergency situation the operator was not able to effectively understand the extent of the upset as the computer screens are not side-by-side. This ended up leading to a catastrophic over-speed event. Compare this to a similar operation where the operator has both computer screens in front of him and a third screen where emergency actions for critical alarms are displayed during upsets.

In *"Recurring Causes of Recent Chemical Accidents"*, Belke describes typical U.S. regulatory treatment of human performance and incident investigation as follows [Belke 1998]:

Regulatory agencies may focus attention on the actions of operators as they reflect the performance of the organization and its management systems. Viewed from this perspective, operator errors, excluding willful negligence or malfeasance, are often symptoms and not really root causes. If an incident investigation program frequently assigns operator error and inadequate training as root causes, or if the recommendations frequently include disciplining operators or conducting more training, this may be a sign that the program isn't identifying or addressing the true root causes. Likewise, if a safety management system relies on properly trained operators to take correct action as the only line of defense against a major disaster, then a facility that employs such a system is asking for trouble in the long run, because humans make mistakes.

This condition has also been termed the error-forcing context (EFC), a situation when particular combinations of performance shaping factors (PSFs) and plant conditions create an environment in which errors are more likely to occur [NRC 2000].

We need to develop organizations where we all more easily recognize and communicate when deviation is occurring. One cultural aspect to achieving improved communication is removing fear from the organization. A culture that requires a guilty party to be identified for every upset or incident does not promote openness and sharing.

Normalization of deviation requires making a decision to violate a process once, or form operating procedures that leave room for interpretation. That is the beginning. Deviation by an operator is typically well intended and not viewed as being a mistake. He or she probably thinks that the change will result in more production, less downtime, fewer steps up and down stairs, or some other benefit. It will not be viewed as doing something negative and definitely not as something that could end up hurting themselves, a coworker, or anyone else. Operating procedures can lack the necessary detail or prescribe a specific way to accomplish a task, so variation can exist between equally skilled operators' techniques. When a deviation is made, and no process related or safety issues are noticed, it may be repeated resulting in normalized deviance.

Deviation by a staff person or manager is no different. The basic driver is some sense of benefit associated with not implementing a process step as planned. In both cases, if there is no consequence (or perceived consequence), the new behavior is likely to be normalized.

Normalization of deviation that results in an incident or near miss is more easily identified. CSB incident reports often directly or indirectly identify normalized deviance as a root cause or contributing factor. Sloppy permitting, inadequate change management, poor housekeeping, and ineffective training are some of the most typical forms of normalized deviation cited.

Think about training. Many companies use the same process and safety training modules every year. Normalization of deviation can develop over time due to apathy and lack of attention. A site using the same safety training and process training quizzes for many years promotes apathy. The quality of the training reflects the value the organization places on operation and process safety training and training in general. Repetitive mandatory safety training can be done well when the instructors keep the sessions fresh with current examples during presentation. Worker training is a huge opportunity for avoiding or quickly responding to normalized deviance.

In the case of BP (Section 2.2.3 below), there were deliberate actions from management that affected process safety such as lack of maintenance, operating without safety systems, and not following recommendations to install a flare, to name just a few. This culture reinforces to the worker that not doing something per the procedure is allowed: management does not follow their own policies when it comes to change management, risk tolerance, and preventive maintenance. Management deviations are far more dangerous operator deviations because: 1) they are broad in scope, and 2) they encourage further normalization of deviation.

The case studies that follow are brief descriptions of incidents in the chemical, refining, and manned space flight industries that all involved normalization of deviation.

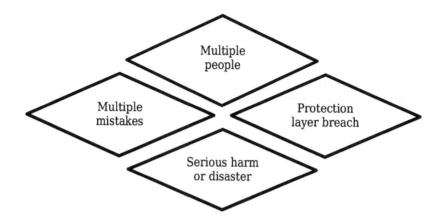

Figure 2.1. Qualifying Characteristics of Normalized Deviance

These catastrophes, and others, illustrate that major incidents have consistently displayed the following four qualifying characteristics shown in Figure 2.1 [Banja 2010]:

1. multiple people
2. committing multiple—often seemingly harmless—mistakes that
3. breach an organization's layers of protection, resulting in
4. serious harm or disaster

2.2.1 Space Shuttle Challenger Incident

In January 1986, the U.S. space shuttle Challenger exploded shortly after liftoff, killing its crew of seven, and causing a severe setback to the U.S. space program. The catastrophic incident was the result of a rubber O-ring seal failure between adjacent sections of the solid fuel rocket boosters. Unprecedented cold weather on the day of the launch reduced the rubber's resiliency, which, combined with the faulty design of the joint, allowed hot combustion gases from the burning rocket to escape. The flames and hot gases burned through the metal supports holding the rocket in position. When the rocket assembly released, it ruptured the side of the external fuel tanks allowing liquid hydrogen and oxygen to mix prematurely and explode.

During the investigation, it became apparent that there was a well-documented history of problems with the rocket booster design including the integrity of the rubber O-ring joints. When the NASA team observed O-ring damage after the second shuttle flight, the they violated internal safety procedures that required them to ground future flights while a cause and solution were identified. Instead the team simply made changes to the assembly process (but not the design) and continued with future flights. The growing momentum to keep the shuttle flying was adversely affecting the team's reaction to the O-ring problem. NASA middle managers repeatedly violated safety rules requiring the prompt resolution of technical problems. Over time, and with 14 of the previous 24 shuttle flights prior to Challenger showing O-ring damage without causing incident to the flight, the managers normalized the deviation, so that it became acceptable and non-deviant to them. Given the success of previous missions with known problems, middle management came to accept the risk and failed to communicate their concerns to top decision makers. While it may have been painful to accept defeat in terms of project schedule, such pain is insignificant when compared to the needless loss of life [NASA 1986].

The NASA organization was and is highly complex. Private consultants were contracted to support important parts of the project and their livelihood depended on the success of the overall mission. Furthermore, the Challenger project operated much like a business with future funding based on past performance. The ability to secure such funding depended on the spreading of good news. Production objectives emphasized the importance of the launch date for each mission. Despite the safety concerns and potential setbacks, work continued on the project. Had the team collectively recommended the suspension of further work until they solved all the technical problems, the catastrophic incident could have been avoided. A strong safety culture would have ensured a diligent and consistent approach to dealing with important issues. The responsibility for making such unpopular decisions in large organizations is often not clear. Effective leadership should have resulted in an organization where team members could step forward and challenge the status quo without fear of consequence.

While the Challenger explosion was not a chemical process incident, the nature of this catastrophic incident was similar to many process safety failures that have occurred in the process industries. In fact, many of the NASA system failures closely align with process safety element failures. The failure of process safety management systems is symptomatic of a failure in the process safety culture of an organization. Process safety culture is the cornerstone upon which an organization can successfully build. Without it, other initiatives will only be partially effective.

In order to identify normalized deviance in an incident, the following questions can be used as a guide:

- What actions might have interrupted the chain of events that led to this incident?
- Why was such action not taken?
- Who had the responsibility to act?
- At what point was the incident inevitable?
- Was there sufficient time to act following identification of an issue?

This incident serves as an example of the type of loss that can occur in a large, complex organization if management systems are not effective, and strong team leadership is not promoted. Additional discussion on the effect of process safety culture and leadership on effective incident investigations is provided in the literature [Klein 2016, Klein 2017].

2.2.2 Space Shuttle Columbia Incident

In February 2003, the U.S. space shuttle Columbia disintegrated during re-entry into the earth's atmosphere killing all seven crewmembers.

The loss of Columbia was a result of damage sustained 81 seconds after launch when a piece of foam insulation the size of a small briefcase broke off the space shuttle external tank (the main propellant tank) under the extreme launch forces. The debris struck the front edge of the left wing, damaging the thermal protection system. The mission continued but ended tragically sixteen days later when the orbiter attempted re-entry. During re-entry of mission STS-107, the damaged area allowed hot gases generated by air resistance against the vehicle to penetrate and destroy the internal wing structure, which caused the in-flight breakup of the vehicle.

NASA's original shuttle design specifications stated that the external tank was not to shed foam or other debris. Potential strikes upon the shuttle itself were safety issues that needed to be resolved before a launch was cleared. Launches were often given the go-ahead as engineers came to see the foam shedding and debris strikes as inevitable and irresolvable, with the rationale that they were either not a threat to safety, or were an acceptable risk. The majority of shuttle launches recorded such foam strikes and thermal tile scarring. It was not until STS-107 that the consequences of this deviation from design specifications were catastrophic.

The Columbia Accident Investigation Board's (CAIB) recommendations addressed both technical and organizational issues. It was established that the failure of the foam insulation had resulted in an impact with a critical wing component. However, the organizational contributors to the incident were far more complex. Among these were normalization of deviance in accepting the debris strikes; denial of vulnerability; ignoring opinions of safety staff; lack of consistent, structured approaches for identifying hazards and assessing risks; and worker intimidation, contributing to communication breakdown.

The Columbia incident was unique in terms of the chain of events that led to the wing failure. However, the management system failures strike a closer resemblance to other major incidents, in particular, the Challenger incident that occurred seventeen years earlier, and the Apollo 1 fire in January of 1967 that killed three astronauts. One of the revealing statements from the investigation report states: "In our view, the NASA organizational culture had as much to do with this accident as the foam." Professor Andrew Hopkins, a renowned sociologist and process safety expert, had this to say about the incident: "[The decisions] were not made by the engineers best equipped to make those decisions but by senior NASA officials who were protected by NASA's bureaucratic structure from the debates about the wisdom of the proposed actions."

Several quality and reliability problems were apparent to staff in the two years prior to the incident. Shedding of foam and debris became common events that were deemed inevitable. Although potential impact incidents on the shuttle were initially viewed as safety concerns that could delay a launch, the tolerance for these increased with each successful launch. The main safety focus appeared to shift to the propellant distribution system that had experienced cracking.

Since the space shuttle program was being operated as a business and had to compete with other programs for federal funding, NASA officials focused on the success of each mission and discounted hazards and risks that were more apparent to technical staff. Technical experts had serious concerns about recurring impact incidents but they did not effectively communicate these to the decision makers. In such a situation, people (management) tend to hear what they want to hear. Normalization of deviance breeds the belief that we are beyond reproach and invulnerable.

The Columbia incident was a great disappointment. Had management engaged in an effective dialogue with workers on the project, it is likely that the true risk might have been perceived and acted upon. Had the recommendations from the Challenger incident in 1986 been fully implemented, the systemic issues that resulted in the loss of Columbia could have been avoided.

2.2.3 BP Texas City Refinery Explosion

On the afternoon of March 23, 2005, an explosion occurred at the BP Texas City Refinery. The CSB deemed it the most serious refinery incident it had investigated to date. The explosion and fire killed 15 people and injured another 180, alarmed the community, and resulted in financial losses exceeding $1.5 billion. The incident occurred during startup of an Isomerization (ISOM) unit, one of the more crucial phases of operation. A raffinate splitter tower was overfilled and pressure relief devices opened. This caused a hot flammable liquid jet to be released from a blowdown stack. The release of highly volatile flammables led to an explosion and devastating fires. The fatalities occurred in or near office trailers located close to the release point. Over 42,000 people were ordered to shelter-in-place and remain indoors. Property damage outside the fence was reported as far away as three-quarters of a mile from the refinery.

The CSB report summarized the situation as follows [CSB 2007]:

> The Baker Panel Report [Baker 2007] found that "significant process safety issues exist at all five U.S. refineries, not just Texas City," and that BP had not instilled "a common unifying process safety culture among its U.S. refineries." The report found "instances of a lack of operating discipline, toleration of serious deviations from safe operating practices, and [that an] apparent complacency toward serious process safety risk existed at each refinery." The Panel concluded that "material deficiencies in process safety performance exist at BP's five U.S. refineries."

The case can be made that the management structure and practices in place during the preceding years provided an environment in which normalized deviance thrived and grew. This culture allowed and supported the following.

- Lack of operating discipline [CCPS 2011]
- Toleration of deviations from safe operating practices
- Complacency toward process safety risks

2.2.4 Toxic Gas Release in Bhopal, India

The 1984 incident in Bhopal, India ranks as the worst chemical facility disaster in history. It claimed the lives of between 3000 to 10,000 people and injured 100,000 others. Many analyses of the Bhopal incident reveal weaknesses in the process safety management programs (many of which did not exist at the time) [Kletz 2009, Bloch 2016].

The incident took place in an intermediate storage area of the facility where highly toxic, liquid methyl isocyanate (MIC) was stored in three separate tanks embedded in a berm. MIC is a feedstock used in the production of a carbamate insecticide. MIC liquid is highly reactive in the presence of water and iron oxide, and the reaction generates heat. Unless removed, this heat will cause a runaway reaction that could release the highly toxic MIC vapor. Therefore, the process design included a refrigeration coil to ensure that the temperature could not exceed 5 °C, a relief system directed to a knock-out pot, a vent gas scrubber, and a flare system to prevent vapor escape.

For several months before the catastrophic incident, conditions at the facility had been deteriorating, with several safety features either shutdown or compromised. Principal factors leading to the magnitude of the gas leak include the following items:

- Storing MIC in large tanks and filling beyond recommended levels
- Poor maintenance
- Failure of several safety systems (such as the tank temperature indicator)
- Safety systems out of service—including the MIC tank refrigeration system, a vent gas scrubber, and a flare, all of which could have mitigated the disaster severity

Other physical contributors to the incident included the following items:

- The MIC tank alarms had not worked for at least four years.
- There was only one manual back-up system, compared to a four-stage system used at facilities operated by the parent company.
- Carbon steel valves were widely used at the facility, even though they corrode when exposed to acid.

Can you tie any of these conditions to possible normalization of deviance?

The incident at Bhopal could have been prevented. There was no real element of surprise. Conditions had deteriorated for many months. Physical and paper evidence suggests there were huge gaps in the safety culture, most likely due to lack of clear operating priorities. A satisfactory process safety culture is a state of excellence that starts with strong leadership and demands discipline and accountability at all levels. This catastrophic incident led to the financial ruin of the parent company, and contributed to global regulatory and executive reviews of process safety protocols [Bloch 2016].

2.3 HOW THE CONCEPT OF NORMALIZATION OF DEVIANCE AFFECTS OVERALL PROCESS SAFETY PERFORMANCE

Management systems are only effective when they are followed. Strict adherence should be expected and required. A total process safety management system should provide simple, interlinked, clearly mandated direction, and guidance for abnormal situations that can be foreseen. Any prescriptive requirements of applicable law plus all internal rules or policies need to be addressed as applicable. Foreseeable situations can be planned for with set guidance or rules-based response.

These documents are analogous to software for humans. They reside within the computer-processing unit that is the organization. In fact, any process safety related documents that give instruction: operating procedures, maintenance procedures, and safe work practices for example, can be considered simply software code for people.

To continue with the software analogy, humans do not necessarily make the most reliable peripherals. Sometimes it isn't the person's fault. The work instructions can be unclear or not up to date. Occasionally the human peripherals can behave erratically and change their own instruction, either through performance error, or through a change or shortcut that can become a normalized deviation.

The one advantage of this process safety software acting through fallible humans is that humans are also capable of observing, responding to, and reprogramming their own software when they find an improvement or a problem in the original instruction. A well-designed PSM system and all its associated procedures will have a way for this user feedback to be processed. All humans filling roles in the process safety system: operators, maintenance workers, safety technicians, engineers, supervisors, managers, and executives should be required to follow their procedures, or modify them through a controlled process when they cannot be followed or can be improved.

The American Institute of Chemical Engineers Center for Chemical Process Safety developed Vision 20/20 which looks into the not-too-distant future to demonstrate what perfect process safety will look like when it is championed by industry [CCPS 2018a]. It is driven by five tenets (Figure 2.2):

1. Committed Culture
2. Vibrant Management Systems
3. Disciplined Adherence to Standards
4. Intentional Competency Development
5. Enhanced Application & Sharing of Lessons Learned

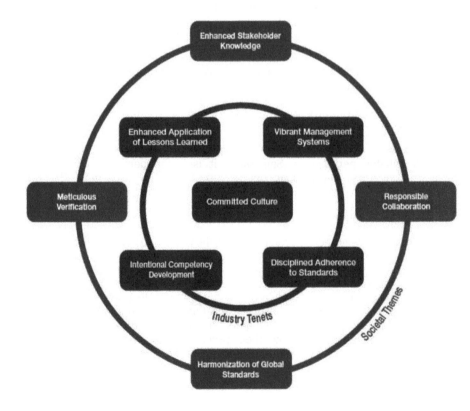

Figure 2.2. CCPS Vision 20/20

These tenets include aspects that help reduce the probability of normalized deviation. Some examples are:

- Organizations need to define expectations for all systems, otherwise it cannot know when deviations occur.
- Systems are rigorously followed, which means all in, every day, every time, thus avoiding deviation. This group behavior can lead to a culture of excellence.

Risk related decision making is made at the proper level of management'

- again checking any deviation against collective knowledge.
- Monitoring existing equipment, age, and maintenance issues is performed to predict failures before critical components wear out. Avoid the *run to failure* attitude. That attitude leads to all sorts of potential for normalized deviation. Work-arounds for broken or underperforming equipment might appear. Parts availability may encourage deviation from the design configuration or in the least cause needless concern about integrity.
- Fit-for-purpose policies and procedures are current; the organization provides the resources to maintain all process documentation. This is a current issue across the industry. Follow your system. If you find it is difficult, change the system to reflect what you actually can do to comply. Evaluate the documents constantly for effectiveness during use and revise according to your document control procedures.
- Reducing normalized deviance and reducing their probability has many business drivers including:
  o Reduced rework
  o Consistent higher quality
  o Less downtime
  o Fewer unplanned outages
  o Faster cycle time
  o Better environmental performance
  o Improved personnel safety
  o Better financial performance

These drivers are the basis of all quality management systems.

2.5 CAN NORMALIZED DEVIATION IN YOUR BUSINESS WORK PROCESSES AFFECT RISK?

Although not often thought about as process safety issues, work processes such as sales, logistics, purchasing, and other business functions can support or detract from process safety performance in a plant. Manufacturing areas can experience infrastructure limitations that may prompt personnel to make exceptions to accepted practices to ensure continued production.

Consider this straightforward example of a deviation driven by logistics and production needs:

> An operating area was notified that delivery interruptions for a key raw material over the Christmas holidays were possible. There was no suitable storage tank available to hold additional deliveries. The decision was made to intentionally fill the tank into the top head to gain additional storage capacity. The field operator had to stand on top of the tank to defeat the level interlock to allow material to fill into the top head. The tank was not designed for liquid above the top weld seam. Intentionally filling the top head with liquid could have resulted in damage to, or failure of, the tank, and loss of containment. How many times had this deviation been previously used to maintain production rates? Once found, the practice was ceased.

The West Fertilizer fire and explosion that occurred on April 17, 2013 in West, Texas, is an example of how normalized deviance in work processes related to sales, production, and logistics increased the risk. Combined with community encroachment over the years, when the explosion occurred, fourteen people were killed, over 160 people injured, and at least 150 buildings were damaged or destroyed [CSB 2016, CCPS 2018b].

Gaps or holes in the U.S. regulatory process were identified that could be considered a form of normalized deviance. For example, although ammonium nitrate (AN) was covered by the OSHA standard *Blasting and Explosive Agents* this was not well known in the fertilizer industry [OSHA 1998, CSB 2016]. Even OSHA inspectors covering fertilizer storage sites did not enforce it. Spring season for a fertilizer plant is a very busy time. The increased inventory ammonium nitrate (NH_4NO_3) and anhydrous ammonia (NH_3) on the site were there to meet the demand. However, the warehousing facility did not take in to account safe storage and handling of the large quantity of the hazardous material. Could this be recognized as a deviation? It is the duty of an organization to seek out and internalize applicable local, state, and federal rules and regulations as a part of their work process? Is not doing so considered normalized deviation? Even when a company has no mission statement, policy, or core value embracing essential process safety principles, including compliance with regulatory requirements, the answer to these questions is yes.

Seemingly, innocuous business processes can be improved by applying the techniques described in this book for specific process safety and organizational normalized deviance. Organizations need to recognize when production and profit goals create a change requiring management. All effective management of change systems need a decision point within the hazard evaluation step where the change can be denied, reconsidered, or possibly reengineered and then resubmitted.

2.6 NORMALIZATION OF DEVIATION AND MANAGEMENT OF CHANGE

The relationship between normalization of deviance and management of change (MOC) is critical. The two are linked numerous times in this book. The MOC and associated pre-startup safety review can be an organization's primary means of managing deviations and preventing normalization.

To compare and contrast the two concepts:
- Both concepts have their basis in human initiated change to a process.
- Effective MOC is focused on managing intentional, planned deviations from the current process configuration or work process in an organized, methodical manner.
- In the alternative, normalized deviance occurs when an organization allows less than fully managed intentional changes to the process configuration or work process. A human has to decide to deviate once and then deviate again and again. After a long enough time, incoming employees may be trained to perform the deviation as if it were the approved work process. This is when intentional changes lead to unintentional consequences. No one doing it will consider it necessary to report it up the management chain.
- Ineffective MOC can cause increased occurrence of normalized deviance.
- Within MOC, an originally well-managed temporary change can turn into a normalized deviation if it remains in place beyond the expiration date agreed to within the MOC system.

Normalization of deviance can exist both internally and externally to the management of change process. A simple flowchart for a management of change system using a request for change (RFC) data form is presented in Figure 2.3 [CCPS 2008a].

Organizations need to recognize unmanaged changes as deviations. Implementing a thorough management of change system, while implementing a risk-based process safety (RBPS) mindset, is essential to protecting work processes from the effects of normalized deviance.

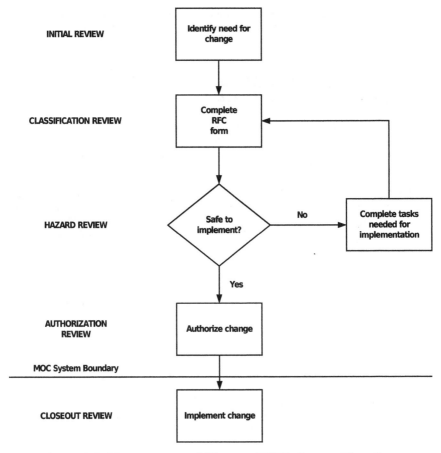

Figure 2.3. Management of Change (MOC) System Flowchart

[CCPS 2008a]

On a broader scale, effective reduction of normalized deviance supports any mature, well-designed process safety and environmental risk management program. It supports operational excellence. It keeps the total program robust and vibrant by avoiding unmanaged changes in business processes. The CCPS booklet *The Business Case for Process Safety* summarizes the benefits of process safety, and these same benefits are supported by actively recognizing and reducing the effects of normalized deviance within your process safety system [CCPS 2006].

Methodically implementing process safety provides four benefits essential to any healthy business. Two of these benefits are qualitative and as a result are somewhat subjective. You can see them in the way the public, your shareholders, government bodies, and your customers relate to your company.

The two remaining benefits are quantitative. These have measurable impact in terms of your bottom line and company performance. All four benefits, when realized together by adhering to a sound process safety system, combine to support the profitability, safety performance, quality, and environmental responsibility of your business.

- **Corporate Responsibility** – Process safety is the embodiment of corporate responsibility and accountability. It helps your company display these characteristics through its actions. The heart of process safety lies in consistently planning to do the right things, then doing them right – consistently. Corporate responsibility leads to the second benefit...
- **Business Flexibility** – Corporate responsibility as demonstrated in your process safety management program leads to a greater range of business flexibility. When you openly display responsibility through implementing an effective process safety program, your company can achieve greater freedom and self-determination.
- **Risk Reduction** – Process safety provides unparalleled loss avoidance capability. A healthy process safety program significantly reduces the risk of catastrophic events and helps prevent the likelihood of human injury, environmental damage, and associated costs that arise from incidents. Although the essence of process safety focuses on preventing catastrophic incidents, the number of less severe incidents is also reduced.
- **Sustained Value** – Process safety relates directly to enhanced shareholder value. When properly implemented, it helps ensure reliable processes that can produce high quality products, on time, and at lower cost. This increases shareholder value."

2.6.1 The Relationship Between Managed Deviation and Unintentional Deviation in Your Work Processes

An unmanaged change is a deviation that starts or continues the normalization process and contributes to the increased risk of hazard exposure. Whether the hazard has a great impact or one of little consequence, the planned work process begins to unravel. The original process is now more disordered.

> A business segment had difficulty locating a suitable warehouse to store oxidizers. A truck full of oxidizers turned up unannounced at a warehouse where hazardous chemicals were never stored. There was no sprinkler system, no city certificate of occupancy to legally allow storage, and residential buildings are within 200 feet of the warehouse. The truck was turned away by the facility management. Through proper management of change, a deviation was prevented.

A better example might be that a particular replacement part is no longer available. Instead of installing a readily available part, a proper review is held to ensure that the part chosen has all of the required features. The change is properly documented.

Changes that the organization plans, controls, and then tracks in a MOC system become well understood. Drawings, procedures, training materials, and other documentation all get updates when needed. However, unintentional changes can still occur.

- **Aging infrastructure of various types.** Steel and concrete equipment support structures deteriorate. Utility systems can be ignored when capital is budgeted for improvements or new equipment is installed to support a business opportunity. Even some critical safety systems can become outdated or unworkable if testing and calibrations are skipped.
    - o For example, a water tube boiler at a chemical plant has a history of tube failures. The inspection report indicates that a total tube replacement is needed due to wall thinning. Funding for the tube replacement is not a priority. A sudden tube rupture near the outer skin of the boiler causes trapped expanding steam to rupture the outer skin and cause major distortion of the boiler drums and tubes assembly.

- **Unintentional undocumented deviations occur.** Workers notice changes that need to be made to the procedures or the P&IDs, but do not report them for various reasons. It might be that no one is assigned as the procedure coordinator whose role it is to evaluate and write, or direct others to write, new procedures, revisions to existing procedures, and remove outdated procedures from worker access. Efforts to increase general site productivity may erroneously classify this as low value-added work.
    - o An example would be a steam turbine generator system that does not have detailed shutdown procedures. This has been noticed but no action has been taken. An operator assumes that the electrical breaker should be opened when there is still some electrical load on the generator. There is an over-speed and subsequent catastrophic failure.

- **Some minor changes or certain types of changes are not considered MOC worthy.** These unconsidered changes that contribute to work process disorder can affect later decisions and risk management considerations.
    - o For example, secondary containment areas for oil-filled transformers have drain valves with a sign on each that they need to be kept shut. Rainwater captured in the secondary containment area is tested to make sure there are no contaminants (such as oil) before draining to the storm-water collection pit. An operator who is assigned to check the secondary containment areas makes a decision to leave the valves in the open position risking containment failure.

3. THE ROOTS OF DEVIATION

I have learned that carelessness and overconfidence are usually far more dangerous than deliberately accepted risks.
Wilbur Wright

Every decision a human or an organization makes involves weighing the associated risks and rewards. Sometimes the analysis is done consciously and conscientiously. When a task is familiar to a person or team, it may be performed unconsciously and automatically, in a conditioned way. In either case, we do not always read the scale properly every time, or we ignore what it is telling us. For example, the farmer is rewarded based on how much food is provided to their family before the growing season is over. Applying this reward system in industry, the shift worker is rewarded based on meeting the daily production targets that contribute to company income. While leadership does not explicitly devalue safety, they may inadvertently do so by not attending to the risk exposure needed to obtain the desired production results. Did the shift worker skip seven steps in an operating procedure to meet production goals? Does leadership even ask, or do they only want to know daily run rates? In situations like this, the desired rewards subsume the perceived cost of the risks.

There is also a difference in an individual's perception of risk at home versus on the job. People may not read instruction manuals at home, may not practice safe lifting or ladder use, even when they are trained in the workplace to identify and mitigate or eliminate those hazards.

3.1 LACK OF OPERATIONAL DISCIPLINE

Operational discipline (OD) is part of the Risk Based Process Safety (RBPS) element, *Conduct of Operations* [CCPS 2007, CCPS 2011]. Conduct of operations, and OD are closely tied to an organization's process safety culture [Klein 2017]. OD is the execution of operational and management tasks in a deliberate and structured manner. Conduct of operations institutionalizes the pursuit of excellence in the performance of every task and minimizes variations in performance. Workers at *every level* are expected to perform their duties with alertness, due thought, full knowledge, sound judgment, and a proper sense of pride and accountability. Some causes of lack of OD are described in the following sections.

3.1.1 Repeated Controlled Deviations Can Lead to Normalization Over Time

What begins as a minor deviation from a rule becomes, over time, the norm. We learn to accept the risk. Having knowledge and skills does not mean a commitment to always doing the correct thing is established. A plant culture that does not require consistent conduct of operations is inviting normalized deviance to occur. Conduct of operation practices include:

- following process design codes and standards
- recording a reading properly and accurately
- following an administrative procedure, operating procedure, maintenance procedure, or safe work practice procedure to the letter
- completing inspection rounds
- issuing work permits to a contractor before work begins
- investigating issues to determine the root cause
- auditing management systems for compliance

Deviations from the practices can occur repeatedly with no noticeable negative outcome. Ultimately, however, each could result in, or contribute to, an incident that could cause serious harm to people and the environment, or have other negative consequences on the organization.

> In one plant operators were required to routinely check telltale pressure gauges for pressure to identify the potential breach of a rupture disk underneath a PSV. The formal check-sheet was available in the control room and regularly completed with check marks to indicate that zero pressure was verified. Field observation during an audit revealed that some pressure gauges had no indicator needles and some had fogged faces that prevented a clear view of the needle position. The routine field checks had clearly become a paperwork exercise conducted from the control room.

When deviations are repeated over time and seem to cause no harm, they may be woven into plant culture and operations. There are any number of conditions that can make deviating from standard procedures and policies seem right. Frequent, controlled deviations may desensitize personnel to the necessity of customary operations. However, the coincidence of imprudence and equipment failure can lead to injury or even disaster. Have you seen things like this occur in your workplace?

- The operators have to leave the control room to hold a manual pump control button during a specific phase of the process, so they tape a rock onto the button and use the process controller transitioning into that step and the high level interlock on the destination to start/stop the charge.
- Due to poorly rationalized alarms, the control system floods the operator with alarms. To avoid the annoyance of nuisance alarms, the operator disconnects the horn from the control system or shorts the acknowledge button.

- For a drum charge operation, the procedure calls for the operator to charge materials A and B, then let the batch mix for 30 minutes before charging material C. To save the steps from the control room to the drum charging area (or due to weather conditions) the operator charges all three materials at the same time.
- Due to the complex nature of the facility management of change system, maintenance technicians routinely begin work on a change prior to all of the sign offs being received for the change, assuming the approvals will occur before the change is ready for startup.
- The workstation and/or control panel design has not been considered during process changes, leading to operators taking shortcuts or having to adjust their activities to account for this.

In a perfect world, there will be troubleshooting procedures or emergency changes. However, when a change is required to respond to an emergency situation (that is, to reduce the risk from an immediate hazard), the change should only be made after notifying the Area Superintendent or Production Superintendent of the change and reviewing the change for its technical basis and impact on safety and health. Such direct support needs to be available. A change authorization form needs to be initiated on or before the next normal business day after notification. Specific emergency change request steps follow.

The benefit of allowing an alternate approval path for emergency changes is that qualified personnel can perform the first hazard evaluation with immediate field access to the change. The change can be approved by available supervisory personnel and subsequently implemented. For emergency changes, a faster approval is allowed so that the change to equipment, chemicals, or technology can be implemented immediately.

Follow-up analysis and reconsideration needs to occur within a set time frame. Typically, the next day the change can be analyzed in full and approved by the proper authority. It can then be made either a permanent change or a temporary change with all action items documented and planned for completion. Maintaining the process safety envelope by using emergency change requests in these situations is a valuable option but should be *extremely* rare.

However, some front line supervisors and operating engineers at a few sites found that they could avoid or delay effort by using an emergency change subroutine when they wanted to avoid the pre-work associated with a standard management of change item for simpler changes. This shortcut in the process was taken advantage of due to the employees' risk versus reward perception. Humans often see the extended analysis and update of the process safety information, operating procedures, maintenance procedures, and safe work practices that may be affected by a change to be unenjoyable tasks compared to efforts to actually move forward with operations or make a quick change to meet a goal, typically production. In addition, an employee may simply have an internal goal to increase the ease of their duties that particular day.

Falsely claiming a change was needed in an emergency mode became normalized within groups of employees. Many sources indicate that once a community at a site or within an organization normalizes a deviant organizational practice, it becomes a routine activity that is commonly anticipated and frequently used [Ashforth 2003, Brief 2001, Ermann 2002, Dunford 2008, and Pinto 2008]. By regularly monitoring the status of the MOC system, the spike in the emergency change category became obvious to the process safety manager. The staff stopped abusing the emergency MOC process once there was clear communication, clarification, and reinforcement of the MOC process. This is an example of an organization recognizing and reducing normalized deviation.

For many years, some manufacturers have used a general deviation request form for documenting various minor modifications to the process—quality, yield, personnel safety, process safety, or regulatory needs. It allows the organization to create records of every deviation to the normal organizational processes that they recognize and want to control. Deviations could then be categorized and tracked. Many companies have kept this deviation request form or deviation approval form concept in place alongside their management of change system to address minor modifications not meeting the criteria for a formal management of change.

One company experienced the same type of misuse with their deviation request forms as previously explained with the emergency change provisions. Employees saw the forms as a way to avoid the more effort-laden management of change system. Again, the employees' risk versus reward systems came into play in making decisions to veer from the approved process. Though they were still using a company approved system, they were using a less rigorous method to manage changes rather than using the required MOC system. Such behavior can be a sign the MOC process is too onerous and needs adjusting.

3.1.2 Accepting Transient Process Conditions as Normal Operations

It is normal to accept the occasional short-term deviation of a condition from a desired process configuration. Does your facility have one or more deviation request forms? There might be a general form as well as others. For example, a form for safety critical equipment related deviations might exist. Who determines what constitutes a deviation and when a deviation is necessary?

Many companies have a policy that anyone in the organization, including employees in operations and maintenance has the authority to shut down the process if they notice an incipient hazardous situation or a safety anomaly. While on paper this is a beneficial policy because the operators and maintenance employees have the most consistent and closest knowledge of the daily operation of the process and are the first to receive indicators that an anomaly is occurring in the process, in actuality, this type of policy should be carefully managed. Sometimes when an anomaly occurs within the process, the operators and maintenance employees will choose a 'ride-it-out' philosophy and will not intervene to shut down the process. The motivation may be "I've seen this before and it will resolve itself in a minute" or "I do not want to get in trouble for shutting the process down if there really was nothing wrong" or even "I know if I shut the process down, it will take a certain number of hours to restart. I do not want to be held responsible for the loss in production." In this case, increasing use of procedures with unambiguous shutdown triggers, and thorough situational training will help the organization fulfill the intent of their policy.

Often chemical facilities have a repetitive need for controlling situations where environmental health and safety critical devices (EHSCDs) are removed from service. If critical instruments require maintenance or calibration, risk increases for the period when they are out of service or bypassed. Other examples of this include fire protection impairments, emergency access impediments such as road closures or traffic restrictions, placement of temporary equipment without proper secondary containment or fixed fire protection, or siting of temporary buildings in an area.

Many sites have an administrative procedure or section of their management of change system that helps ensure effective analysis, authorization, communication, and recordkeeping for disabling environmental health and safety critical devices for brief out-of-service periods. That method can always be used for managing EHSCD shutdowns if desired. The steps described in such a procedure can provide an effective and streamlined approach for managing these specific types of changes. A full management of change review is generally required if the outage goes beyond a few days. Often the procedure includes a table of acceptable mitigation measures as minimum guidance for the staff to add appropriate layers of protection during instrument outages [CCPS 2008a].

Process safety professionals should ask stakeholders regularly if there are improvements that can be made to the MOC process. Any separate approved deviation management system should be aligned to the facility's overarching management of change system. Special procedures describing a deviation should be written, reviewed, and approved prior to the transient process condition. It should include a process for ending deviance once a transient process condition is over as well as a check to ensure that normal operations have resumed. For example, the commercial nuclear power industry utilizes quantitative risk assessment to identify risk critical components and monitors them to ensure their downtimes are within set allowed outage time (AOT) limitations. Management of change is an excellent risk reduction method to manage deviance and provides long-term rewards when implemented rigorously.

3.1.3 What About Your Infrastructure?

The experience a person has of buying a new vehicle and keeping it for the duration of its useful life captures the phenomenon of getting used to aging infrastructure. Small deficiencies such as a window that sticks or a dash knob that is missing are signs of minor corrective maintenance deviations. They may be lived with or fixed in due time in accordance with the vehicle owner's philosophical position on run-to-failure maintenance practices. However, even the most beloved and trustworthy vehicle eventually ages out of useful service. Failing reliability and expensive repairs may be the final decision point in replacing the vehicle, but safety will almost certainly be enhanced merely by getting a newer model.

Aging Infrastructure

A chemical processing plant, refinery, or any industrial facility is like that car. The drivers of both can get used to their quirks in operation and handling. The difference is that the owner of the plant should have a goal to be a high reliability organization (HRO) [van Stralen 2018]. An HRO should make a special effort to adopt a maintenance mindset more advanced than run to failure. Technological advancements across many fields of computer, process equipment, and structural engineering have made some infrastructure improvements a business necessity for cost competitive reasons.

Another aspect of this issue is that people often do not consciously consider slow, progressive aging or natural wear and tear on infrastructure as deviation. That is, until there is an infrastructure failure. In some cases, it happens so slowly that the equipment's changes are unnoticeable without testing programs of some sort. Business practices like changes of ownership or delaying maintenance due to budget considerations (and even delaying maintenance turnarounds due to extremely profitable market conditions) might push the limits of an aging processing unit. It could have been ignored so long that its real physical engineering parameters no longer match its design. Realize also that the catastrophic failure may occur in one small aged or worn piece of equipment while the bulk of the process equipment is serviceable. People may tolerate outages that are more frequent or bypasses to keep the facility running and maintain production. *Dealing with Aging Process Facilities and Infrastructure* deals with this subject in more detail [CCPS 2018c].

Employees can actually become used to the quirks of the aging equipment and controls. Knowing the nuances of an old process, such as a check valve that needs a whack during startup, may actually reinforce a sense of mastery. In addition, when new equipment replaces the old, it can work differently, produce better quality, or be more efficient, but the learning curve to operate it can lead to deviations. Two examples follow.

- A railcar-loading pipe rack at one site contained toxic chemicals. The painting budget had been cut several times over a period of 15 years. Active corrosion was apparent. No actions were taken until the pipe rack support failed and there was partial collapse of the pipe rack.

- Pipe rack I-beams adjacent to furnaces support piping systems containing flammable and toxic liquids. Corrosion under the fireproofing occurred to the extent that there were corrosion holes several inches in diameter in the supports. Deferring of maintenance had been normalized. Consideration had not been given to the remaining fireproofed I-beams' adequacy and the potential for fire escalation in the event of a furnace fire.

Third party audits by insurance groups or consulting firms can help to see an aging facility with fresh eyes at regular intervals. They are often helpful in influencing equipment upgrades that reduce risk.

Incremental Change over Time

Improvements and incremental change may cause deviations. Many small changes over time can add up to big changes. Some group or position in the organization should own this process. By only focusing on discrete decisions or small incremental changes, the organization can to fail to see the big picture.

When is it time to invest in capital improvement? When is enough truly enough for some units or pieces of equipment? Keeping older equipment may mean that the site cannot get original parts anymore. We overlook the chance to buy new or different equipment that may work better or is more environmentally sound. Question how well your plant managed the changes over years that were involved in moving from a panel board to distributed control system (DCS), and then adding safety instrumented systems and programmable logic controllers. Did we really manage this change effectively?

3.1.4 Examples of Practices That You May Get Away With...Until You Don't

One way to begin the identification process is to look at deviations that are typical across your industry and explore whether they are occurring at your site. A list of examples is below:

- A blown rupture disk on a near-atmospheric vessel allowed to vent until it is convenient to replace it. With each occurrence, the venting time gets longer.
- Dike drains routinely left open after rain water is drained out.
- Vibration from rotating equipment stressing piping (in one plant the vibration was not known by management until a few days before a customer audit, at which time they thought it would reflect poorly on them. Ironically, once a bad bearing was replaced and the rotor rebalanced, the productivity of the equipment doubled).
- Numerous plugs and caps missing in the plant, not replaced after a vent, drain or clean-out port is used.
- A work-around (with or without a temporary MOC) used repeatedly rather than addressing the underlying cause. The problem is repeatedly fixed rather than eliminated.
- A leaking flange repeatedly tightened rather than investigating the cause for the recurring leak (such as an improper gasket).
- A safety instrumented system (SIS) left out of service because of repeated issues with operability.

- A temporary modification such as a jumper hose left in place after a nuisance trip, or "in case it is needed again."
- Personnel taking a short cut through a restricted area to save time/walking in inclement weather.
- Completing a field log sheet from the control room because *there is never an issue.*
- Approving a work permit without conducting a field inspection because "this contractor knows what he is doing."
- Condensate dripping from cold lines or equipment not addressed because it is just water. A problem later develops due to corrosion on equipment or lines below.
- Normalized existence of water drips. (It may not be water the next time).
- Operating with less than a full staff for the shift.
- Ignoring alarms because they are always sounding and are a nuisance.
- Lack of housekeeping leaves potentially flammable materials in the operating area.

3.2 INSUFFICIENT KNOWLEDGE, PROCEDURES, TRAINING AND RESOURCES

Process Knowledge Management, Operating Procedures, and *Training and Performance Assurance,* are three of the 20 RBPS elements [CCPS 2007]. These elements are interlinked. Collection and dissemination of process safety knowledge is essential to creating good operating procedures and training modules. Gaps in knowledge can lead to bad decisions. Incident investigations sometimes find that knowledge of the process hazard and how the process deviation could create the hazard was known by someone in the organization, but not by the people who had to make the decision of how to respond to the deviation. In other cases the operating procedures or process training do not contain the hazards of deviations, which can enable a lack of OD, since people can be unaware of the hazards.

3.2.1 Consider Loss of Organizational Knowledge

To quote the late process safety expert and author Trevor Kletz "organizations have no memory, only people do [Kletz 1993]." By this, he means that an organization needs to maintain a repository of lessons learned and revisit it regularly when evaluating new and routine situations. Otherwise, the past behaviors that contributed to the incident can easily creep back into the organization. The organization should document its memory in accurate work processes and process information.

To combat organizational memory loss, one chemical corporation recognizes the importance of previous incidents and the lessons learned. It develops a program where select incidents are assigned an anniversary date.

Communications are sent out company-wide on the anniversary dates. Another organization uses *critical knowledge training modules* to spread incident learning. In lieu of sending written communications, their staff develops brief video awareness-training modules, presented by individuals that were affected by a specific incident. It is mandatory training and tracked in an electronic learning management system. The U.S. Nuclear Regulatory Commission publishes Knowledge Management Digest pamphlets of around 20 pages summarizing major incidents and lessons learned. A DVD with video history and relevant references is also included.

Many manufacturing companies are faced with valuable knowledge and skills literally walking out of their facilities. A wave of retirement is occurring as workers age out of the workforce. Some companies are so shorthanded they do not take the time to ensure that this group of workers has a chance to transfer their experience in some way before they leave the company. These experienced employees can continue contributing to the organization's future by upgrading the operating procedures and training materials as they wind down their tenure or by returning as part-time instructors for new hires. The examples for evaluating procedures and training provided previously in this book apply to this situation as well. When organizations expect employees to be smart, they are obligated to make access to information a primary business concern. The challenge of transferring valuable knowledge is even more complex for large global organizations. Historical incidents that shaped a company's standards, procedures and policies may have occurred prior to becoming a global organization. That means the most intense communication was limited to the company's home country or region of origin. Access to knowledge is a significant human factors issue. Serious near misses may not be known beyond the plant or unit where they occurred. Review incidents with the entire organization and even share within your industry through organizations like CCPS, the American Petroleum Institute (API), and others.

The tendency to outsource aspects of maintenance, shipping, warehousing, and other traditionally in-house functions causes a special problem. The organization has put its knowledge base at risk. Yes, some facilities maintain the same basic roster of personnel even when a contracting company changes, but contract workers often come and go with opportunity for better wages or work conditions. When the workers most exposed to the day-to-day operations of the equipment are changeable and bring ingrained work culture issues and lack of experience, deviations can increase.

Anecdotal experience holds that a well-designed and maintained process safety management system at a site can be maintained for 10 to 15 years at a high quality level under one corporate owner, but programs can deteriorate drastically after one or two changes of ownership. Even reorganizations of an internal nature cause disruption of day-to-day work processes. Deviations can occur as employees cope with the disruption as a way to keep things going.

Examine Excessive Personnel Changes

Excessive personnel changes introduce confusion into an organization. *Frequent organizational change* happens to be one noted catastrophic incident warning sign. Warning signs are indicators that something is wrong or about to go wrong. It is common to see cases where there are frequent movements of key people due to promotion or other reasons. Appendix A – *A Survey To Help Identify Warning Signs Of Deviations* provides the list of warning signs and a detailed survey tool developed from the book *Recognizing Catastrophic Incident Warning Signs in the Process Industries* [CCPS 2012]. *Guidelines for Managing Process Safety Risks During Organization Change* provides a process for managing temporary and permanent organizational changes [CCPS 2013].

Your plant operation can suffer when employees who fill key roles are not fully prepared for taking over new duties without supervision or mentoring fills several key positions. When the organization has no written mentoring or backfill plan, there is reluctance to use two people on the same job as part of the transition to train the newer hire. Questions to ask include:

- Do your job descriptions have minimum qualifications?
- Do you test for proficiency in the skills required to do the job?
- How do you assess an employee's skill at risk acceptance, identification, and understanding within your corporate risk criteria?

Each new person might be unwittingly introducing deviations into the work processes through the act of learning a job on the fly. People at all levels of an organization need to be able take time to think to perform at their best. Job design and staffing for normal operation means your staff could be too lean in experience when upsets occur. For example, some organizations seem to have an unwritten rule that workers are valued for multitasking. In fact, the idea of multi-tasking is a myth. In most cases people are not multitasking but switching between tasks rapidly.

One tool that can address questions concerning personnel and organization changes is a competency matrix as described in *Guidelines for Defining Process Safety Competency Requirements* [CCPS 2015a]. The matrix lists the process safety skills and responsibilities of positions throughout the organization. It enables companies to determine training requirements and manage the effect of personnel changes throughout the organization.

3.2.2 Check That the Training Program Matches Current Process and Job Configuration

Interviews with operations and maintenance personnel can reveal their opinions about their training program's effectiveness and indicate where to look for deviations. Comments may be that the training materials are outdated and do not match the plant design. This is a sign of normalized deviation. Management of change is the element of the system where these training process deviations began. The ease of documenting computer based training (CBT) user activity and test scores is attractive to organizations. The effort required to develop effective training and maintain it as current is not. Avoid making the tests so easy to pass that it obviously biases the results. Always test to the well-written objectives of the training modules. At some plants where these recommendations are not heeded, operators have reported they had to answer computer based training module test questions incorrectly in order to pass a required test.

3.2.3 Evaluate Periods of Resource Constraints

Whether the cause of the constraints arise from the market or from within the organization, deviation can be encouraged as workers attempt to work with less.

Head count restrictions, hiring freezes, and the empty seat syndrome all restrict available human resources. Empty seat syndrome is a when a position is not filled because of a lack of qualified candidates, a slow hiring process, or the economic appeal of not paying that salary and having the position's critical tasks performed by others for a few months, or years. Operations workers report that, when they are given field promotions of this sort due to human resource shortages, they feel un-empowered and on edge in their expanded position. Not preparing your workforce for their new duties and tasks can cause them to feel overwhelmed by the new tasks loaded onto them for an undetermined amount of time. It drives people to search for short-cuts to get the work done and can create a mindset that leads to deviation. As noted in Section 3.2.2 a process safety competency matrix as described in *Guidelines for Defining Process Safety Competency Requirements* can provide a rational way to manage these issues [CCPS 2015a].

Material resource constraints during cost cutting efforts can be needed to ensure the long-term success of the organization. Sometimes the cuts are to increase short-term profits, or in response to economic conditions. Again, the projected mindset of the organization onto employees may allow the hourly workforce, technical staff, and leadership to infer that shortcuts that cut costs are welcomed. The 2015 Volkswagen emissions scandal, where a well-respected global automaker displayed direct disregard for regulations and accepted societal norms, is a case worth studying [Hotten 2015]. This case may be an example of a normalized deviation driven by a cost cutting or profit seeking mindset and resulting emphasis at the organizational level.

3.2.4 Failure to Document the Infrastructure Knowledge

Inadequate or missing documentation of infrastructure is a common issue at older sites. This can be related to roads, pump and equipment pads, dikes, etc. and particularly to underground piping, electrical cables and waste lines. There may be only one senior person left on staff to be consulted when this information is needed.

> Lack of Operational Discipline (OD) when finishing a project in the mid 70's resulted in temporary power and phone lines being abandoned in-place. A decade later when that area was being excavated for a new project, work was slowed by the need to test every line as it was uncovered.

Inaccurate documentation of infrastructure presents another form of normalization of deviance. Frequently during a PHA, some drawings are found to be inaccurate. There may be a recommendation to update those, but several occurrences of these may indicate other potential gaps in infrastructure documentation. In this case a recommendation to review such safety information and infrastructure knowledge may be called for.

While time-consuming, in the long run it is worth the effort to plan a project to locate and assess infrastructure, then address the issues found in a prioritized manner.

3.2.5 Human Factors Issues

Human factors is the practice of designing products, systems, and processes that enhance the ease and reliability of human interaction when they are used. Traditional views of human factors and ergonomics bring to mind things like desk and chair height. Physical aspects in a plant, like the ones called out in Table 3.1 should be considered also. Beyond ergonomics, human factors and human reliability analysis (HRA) strive to identify the cognitive processes common to a wide variety of error types [Reason 1990].

We often forget that the documents and software systems we use have information transfer related human factors aspects. How do these factors contribute to the human error that can encourage deviation? *Subliminal: How your unconscious mind rules your behavior* describes one human factors aspect of information delivery, the fluency effect [Mlodinow 2012]. The fluency effect describes the phenomenon arising from the fact that the form in which information is presented affects our judgments about the substance of that information. That is, accurate information presented in an unappealing or hard to access format is less trusted than inaccurate information that is presented in an appealing way.

One common example of the fluency effect many people experience is when performing a web search. A poorly designed website or software interface might have just the information you need in it, but its lack of visual appeal and difficulty of use may put you off. This first impression can send the searcher seeking another source, or it can influence them to simply guess at the result they need.

Computer hackers take advantage of the fluency effect in making pop-up windows that look like legitimate trusted images one has clicked on before. The effect applies to much more than just documents. Operators cite DCS control system screen design as a human factors issue. Alarm system design is another issue cited. The fluency effect essentially applies to documents, drawings, computerized visual representations, and any other sensory input that humans need to reference to operate a chemical processing facility.

Table 3.1 Traditional Human Factors Issues

Control room design
Tool design
Labeling of equipment, piping, critical valves, field instruments, switches
Location of manual valves
Location of field instruments
Location of sampling points
Operator task safety
Operator task ergonomics
Opportunities for operator error
Non-routine tasks
Chemical Exposure Hazards
Potential exposures
Adequate engineering controls
Building ventilation/fresh air intakes
Toxic gas monitors, alarms
Protective equipment location
Placarding

Organizations need to make sure all systems are as intuitive and well suited to an employee's successful physical and cognitive use. This shows the workforce that the organization values their compliance with the group's work processes. Examples of systems that have human factors interfaces for operations and sometimes maintenance are listed below:

- Process control instrumentation
- Process vessels and piping
- Process software, especially the operator interface
- Printed or electronically delivered information packages such as procedures and P&IDs

Quality improvements are primarily sought by improved process techniques and materials. However, a holistic approach toward quality assurance requires that due consideration be given to improving operator efficiency through human factors. The areas to address are

- job design,
- equipment design,
- man/machine interface design,
- personal interaction,
- organizational structure, and
- work environment.

The AIChE Academy makes available numerous conference presentations on the topic of human factors that provide experience and insights based on process incident lessons learned [AIChE 2018].

3.2.6 Insourcing Jobs, Outsourcing Jobs, and Normalized Deviance

Insourcing is doing tasks that used to be performed by a contractor or other non-direct employee in-house, or with internal resources. Outsourcing is the opposite, having tasks that used to be done in-house assigned to contractors. Normalized deviance can increase in either situation without proper decision-making and preplanning. Consider the following two questions if your organization is considering either method of accomplishing critical work:

- Do we have the internal competencies and worker capacities needed to insource the work without affecting performance of existing day-to-day duties?
- Has the organization analyzed the outsourced function, prequalified the contractor, and provided the contractor with all of the information and training needed to perform the work effectively so that cost savings are maximized while adequately controlling risk?

If your organization does choose to outsource one or more critical tasks, good *Contractor Management* is essential. A good practice is to have subject matter experts (SME) within your company review contract work. As with all RBPS elements, periodically audit the work quality and adherence to work processes required of your service providers.

Examples of problems companies have encountered include:

- Technical training and procedures prepared by contractors who have never worked in the chemical processing industry in the functional area under development or who do not have access to your subject matter experts.

- Contractors unaware of the site's safety culture, paradigms, specific nomenclature, and jargon.
- The fluency effect, i.e., beautiful process safety and work process documents that are technically inaccurate or not written to the level of detail needed, offering style over substance rather than style and substance.

Beware of assuming that outside experts have something special that your internal employees do not have. Learn how to hire outside expertise your organization does not have enough of or does not have at all. Then inspect what you expect in accordance with the contract to ensure the quality of the work product.

3.3.7 Organizational Change can Encourage Normalized Deviance

Has your job ever changed even though the title, and the pay, did not? Do you frequently hear the phrase "work smarter, not harder"? Introducing new technologies and work responsibilities to a position can cause an employee to be less concerned with strict adherence to standards. A *do more with less* attitude in the organization can lead to trying to do the necessary work with too few people.

While the first tasks to be lost may truly be low value-added work, at some point the organization is faced with cutting essential work due to the lack of man-hours to accomplish this work. This can be a very big problem when management makes staffing decisions without appropriate knowledge of the true minimum staffing level (and associated skill sets and experience) that are necessary to adequately do the job. The trend of early retirement for experienced personnel and replacing them with recent college graduates is an example, as are efforts to reduce labor by X% with across-the-board cuts.

A related organizational change problem is position elimination requiring an employee to have to do two or three jobs without adequate transition periods, training, or time to get them done. When left unimpeded, individual deviance leads to institutional deviance. It is an obligation of the institution to curtail this progression. The industry and regulatory agency awareness of and interest in organizational management of change indicates widespread recognition that staffing levels and employee training and development at all levels of the organization affect process safety performance. Future audits and inspections may commonly address whether sites are staffed—or have access to needed resources—for process safety compliance.

Once again, as noted in Section 3.2.2, a process safety competency matrix described in *Guidelines for Defining Process Safety Competency Requirements* can provide a rational way to manage these issues [CCPS 2015a].

3.3 RISK VERSUS REWARD PERCEPTION

3.3.1 How Employees Perceive Leadership's Message

When the organization communicates its needs to reduce cost or increase productivity, these messages can be interpreted by operations and staff personnel as a need to take short cuts, push operating limits, or challenge themselves. This is particularly true when goals are not well thought out. When employees collectively develop the perception that management is encouraging a *by any means necessary* attitude and approach to achieve stated objectives, normalized deviance can be inadvertently reinforced. Rewards for meeting production goals, being innovative, or saving money can even stimulate this mindset in certain situations and organizational cultures.

Past Reward for Deviation Encourages Deviation

There are two types of rewards for normalized deviance. One is the immediate reward of taking that shortcut to save personal or organizational effort or the reward of increasing yield but, at the same time, increasing risk exposure by deviating. Deviation may even result in monetary reward for an individual or team as described above. The short-term gain, even accumulated over time, can be negated by one incident.

> Over several years, an employee had bypassed an interlock to clear jammed material on a conveyor to save 15 seconds. Finally, when trying to clear a jam, the machine activated and broke her wrist. The machine was down for 12 hours while the entire control program was reviewed and all interlocks retested.

The other type of reward is when a deviation actually reveals a need to modify a process everyone had accepted as optimal or a normalized deviation improves a key performance indicator. Maintain awareness of when these conditions may be present in the workforce.

Experiencing rare positive outcomes from deviations never justifies sloppiness or accepted error-creep in operations. Normalized deviation needs to be viewed as unacceptable to drive the process safety culture away from increased risk taking. If an organization has serendipitously positive results more frequently than it experiences the incidents and catastrophic events (those with the 1,000 and 10,000-year likelihoods), this may lead managers to downplay the problems resulting from normalization of deviation. The seriousness of its potential impact is lost. Ask any witness to a catastrophic process safety incident how they were affected. Such events cannot be allowed to happen.

Does the Organization Accept Deviation?

Be alert for situations where the organization is accepting and not correcting deviations because they have not yet experienced adverse consequences. One example is when a major goal has been met due to deviating from normal work processes. Management provides positive feedback and possibly rewards the team—maybe unknowingly reinforcing deviation. Driven to progress to the next step or phase of a project, leaders may not think to ask, *how did you do that?*

Given the low frequency of catastrophic events, positive or neutral results from deviations can be overlooked or not believed. Managers are often presented general information without sufficient detail for them to truly assess the risk. Leadership is often inclined to skim over the issue or fit its assessment into their busy schedule. When they talk to each other, it is at a high level, that is, in generalities.

Look for Inherent Weakness in Unilateral Business Decisions

An inherent weakness is a hidden defect or nature of a thing, system, or a decision that causes or is a contributing factor to its decline, impairment, or ultimate depletion. Examine every leadership decision against its inherent weaknesses. Prepare to address the cost of accepting these defects in the equipment, systems, or specific leadership decisions. In some cases, the well-known inherent weakness of an activity or good creates an unacceptable risk ranking for a carrier or insurer. When a characteristic or defect is not perceptible, and if the carrier or insurer has not been warned of it, neither may be liable for claims due to the hidden defect.

Take the time to evaluate executive and leadership team decisions against your corporate values and mission statement. Is the risk acceptable? Can it be reduced so that it achieves suitable risk level? Determine the specific presence, and acceptability levels of risk, and the layers of protection proper for the inherent vice in these organizational decisions and ensuing behaviors when it is unavoidable.

3.3.2 Evaluate Management Systems

Evaluate your process safety related management systems. The best working systems are those designed to: 1) match the organizational culture and resources of the site and 2) connect the elements that interact with each other for optimal ease of use and understanding to look for evidence of normalized deviation and potential sources.

Your management system may consist of the following elements in a program based on the model presented in Chapter 2:

- **Employee Participation.** The commitments made in this document on how to involve employees at all levels of the organization in process safety can be checked to verify compliance.

- **Process Safety Information.** Accurate process safety information can be checked by verifying the latest versions of drawings and documents against the current process configuration.

- **Hazard Identification and Risk Analysis Program.** Evaluate the quality of the PHA and documentation of action items against open or rescheduled status. A checklist for evaluating a PHA against essential criteria is presented in Appendix B of the CCPS publication *Revalidating Process Hazard Analysis* [CCPS 2001].

- **Procedures Program.** Evaluate the accuracy of the operating procedures against the current process configuration. Thorough reflection of the plant configuration in the P&IDs, PFDs, operating procedures, and all PSI is the goal. Discrepancies found here indicate failures in MOC/PSSR elements and lapses in other process safety elements that cascade from unmanaged change. If procedure accuracy lapses enough and the plant is producing product, the procedures are not being followed. This is an example of normalized deviance at an institutional level. This also applies to Safe Work Practices, and Emergency Response Procedures. Maintenance procedures can be addressed by this program or the Mechanical Integrity and Reliability program.

- **Training and Performance Assurance Program.** Evaluate the accuracy of the process overview training modules used for training all employees against the current process configuration. Thorough update and consistent information is the goal. Deviations here also indicate failures in MOC/PSSR and lapses in multiple other process safety elements. Some sites regularly qualify operators on training modules that are outdated and procedures that do not work. One suggestion is to involve the people being trained in helping to identify deviations. Involve them in finding where the training materials and procedures do not match how they operate the process.

- **Contractor Management Program.** Evaluate both the quality performance and safety performance of critical contractors. Check the organization's written commitments to contractors concerning process safety and personnel safety training and documentation.

- **Operational Readiness Program.** Evaluate randomly selected PSSRs and trace their steps through the work process with a strict evaluation against the current process configuration. A PSSR is a second check against the MOC process prior to startup. A failure here can allow deviations in the MOC process to propagate. More importantly, failed PSSR can also allow units and process changes to be started up without the benefit of a full evaluation of the risk. Incidents or near-misses that occur during check-out and start-up can be indicators of missed items in PSSR.

- **Mechanical Integrity/Asset Integrity and Reliability Program.** Evaluate samples of all maintenance related documentation such as tests and calibrations, training records and module, materials of construction data, maintenance procedures, mechanical integrity system procedures, and others. Consistent implementation of the mechanical integrity program is the goal. Look for deviations and delayed maintenance action items on critical equipment.

- **Safe Work Permit Program.** A common deviation in this long-standing safe work practice is due to normalized deviance. Workers may complete so many of these that a crew sometimes unanimously agrees to sign off improperly inspected and evaluated hot work activities. The round-the-clock heavy maintenance load of a plant refurbishment (turnaround) is an example of when this can occur. Find deviations by field auditing and observing permit execution and the hot work.

- **Management of Change.** Like *Operational Readiness*, review randomly selected MOC reviews and trace their steps through the work process with a strict evaluation against the process. Allowing failures in the MOC process to persist allows deviations in the entire process safety program to proliferate. Deviations in MOC are responsible for the eventual failure of many originally well-designed and implemented process safety programs. Sites that have had multiple owners in their history are especially vulnerable to deviations. Review random work orders to determine if a change was implemented without an MOC.

- **Incident Investigation.** The investigation provides an opportunity (sadly, after the fact) to look for normalization of deviation. A method to look for organizational deviation is to evaluate the action item implementation. Slow response to process safety incidents indicates a deviation from the industry good practice, and often company policy, of enhancing organizational learning from studying failure. Data analysis of incidents, especially with an eye for repeat causes is an effective deterrent to normalization of deviation.

- **Emergency Management.** Evaluate the drills and drill critiques. Ensure the drills themselves are rigorous enough in their design to match the level of catastrophic release potential for the site.

> An ammonia refrigeration unit developed a liquid leak when a plug came loose from a valve. The refrigeration skid had a fire protection deluge sprinkler misting system designed to mitigate an ammonia cloud. During the leak, the emergency responders did not activate the water-spray system. The emergency response procedure did not contain any details about the spray mitigation system and the responders were unaware of its existence. Lack of detail in equipment specific emergency response procedures had been normalized.

- **Auditing.** Evaluate the most recent periodic reports against reality. Interviewing the hourly workers can verify discrepancies between program and practice or between documented and actual performance. This deviance can be found both when in-house personnel perform the audits and when a third party consultant performs the audits.

- **Risk Based Process Safety Guidelines.** This document describes non-mandatory guidelines for assisting with continual enhancement of a site's management system to match the culture and resources of the organization. Using risk based techniques can make it more likely that workers will follow the desired process and not deviate. The objective of risk based reduction is proportional effort for low risk items and increased attention placed on higher risk items.

3.3.3 Incentive Methods

Whether an incentive is given for production and profit or safety goals, there is a potential negative side to rewarding positive outcomes. Extensive clinical research examining situational observations on motivation and incentivization has been performed. They involve two categories of method:

- appetitive methods (receiving a positive reward)
- aversive methods (avoiding punishment)

Balsam and Bondy, researchers at Barnard College and Rutgers University respectively, found that at best, the negative side effects of reward are real and individual response to reward of any type is unpredictable [Balsam 2003]. For example, use of rewards may cause cheating. Applied to safety and process safety, this means rewards for reductions in injury or incident rates may lead to under reporting.

The basic psychology of motivation is well studied and can help in understanding what makes people deviate or comply. The book *Drive: The surprising truth about what motivates us* explores the past 50 years of motivational research [Pink 2009]. The book has one chapter titled *Seven Reasons Carrots and Sticks Don't Work* and another chapter titled *...and the Special Circumstances When They Do*. This emphasizes the issues organizations face regarding incentivization and using it to control deviation. Pink's work finds that three primary factors combine to motivate individuals toward higher performance in work, school, and life: autonomy, mastery, and purpose.

- **Autonomy** – having some measure of control over your actions and work
- **Mastery** – having the sense that you are, or are able to be, good at what you are doing
- **Purpose** – holding a sense of some self-determined higher purpose related to the goal

Can these three things come together to motivate an employee to deviate? Certainly. Recall the three mindsets described in Chapter 2. A worker can be seeking motivation through these three means but still be unknowing of the hazard, unbelieving of the probability, and underestimating of the severity of the outcome of a deviation. They may feel making the deviation supports their display of one of the three elements of motivation Pink describes. Few incentivization programs consider all three factors. Evaluate your facility's approach to reward [Pink 2009].

3.3.4 A Low-value Work Perception?

An organization naturally wants to focus its efforts on the work processes that will have the biggest impact on business results. Achieving this impact involves identifying and cutting out low-value work. But how does the organization know the true value of a task?

Deviations can occur when individuals decide that some of their tasks are of little value or even think that their entire position is a low-value work activity. A non-motivated worker, one who has lost a sense of purpose and engagement, may take shortcuts or delay action. Housekeeping is a classic example. If you have ever performed any type of workplace audit, simply walking into a poorly kept work area can give you a general impression that deviation is likely. When an entire facility seems unloved, it is usually a warning sign that there might be process safety lapses. A chemical safety board (CSB) investigation of the Port Wentworth Imperial Sugar Company dust explosion and fire cited lack of housekeeping coupled with workers not understanding the process safety hazard of the fine sugar dust was at the root of that event [CSB 2009].

Occasionally, an organization's management decides that tasks are low-value without understanding the true nature of the task. While this may give the appearance of increasing shareholder value, management should consider making better use of their technical staff and observational skills before making these sweeping decisions that can establish an organizational mindset.

3.3.5 A Culture Biased Toward Action Rather Than Analysis?

Take a visit to some manufacturing sites and spend time observing the workforce at various levels of the organization. One can leave a site with the sense that it is not acceptable to take the time needed to get things done in a methodical fashion. It seems the organization's emphasis is placed on speed of response and assignment of blame rather than the holistic quality approach that allows an organization to make continuous gradual progress across all its business needs through analysis and appropriately applied rapidity of action.

Festina Lente is a Latin motto that translates into the English phrase *make haste slowly*. The message here is that plans should be made and activities should be performed with a proper balance of urgency and diligence. If tasks are rushed, then mistakes are made and good long-term results are not achieved. Work is best done in a state of flow in which one is fully engaged by the task. All of us know this state is difficult to achieve but we have probably felt it at some point. Slow is smooth, smooth is fast is a version of this message and is used by elite military teams from countries around the world to help them embrace the best mindset to achieve their assigned missions. Using this approach in organizational planning and implementation can assist in preventing deviation.

3.4 OVERCONFIDENCE

3.4.1 The Dangers of Overconfidence

Wilbur Wright, a father of aviation, knew that overconfidence can be treacherous, as realized in the quote that opens this chapter. Creeping and infectious, overconfidence may lead one to discount risks and overly rely on redundant systems, passive mitigation systems or worse yet, luck. In an instant, overconfidence by an individual or a team can completely offset any preplanning an organization has established for recognizing and addressing deviations in a logical method.

What causes overconfidence in the face of risk? In some of us, it might be due to a personality typology and will exist over the long term. Some short-term cases may be due to instances of hubris, or what is termed the Dunning-Kruger effect.

The Dunning-Kruger Effect

Dunning and Kruger of Cornell University and the University of Illinois, won the Ig$^{®}$ Nobel prize in psychology for their report, "Unskilled and Unaware of It: How Difficulties in Recognizing One's Own Incompetence Lead to Inflated Self-Assessments" [Kruger 1999].

In a highly awaited award ceremony each fall at Harvard University, the Ig$^{®}$ Nobel Prizes fulfill their purpose to "honor achievements that make people laugh, and then think. The prizes are intended to celebrate the unusual, honor the imaginative — and spur people's interest in science, medicine, and technology." All Ig® Nobel winners have produced science, but the award often points out some aspect of folly or obviousness related to the work. However, after generating a few laughs at the results, Dunning and Kruger found their results were repeated by several other researchers over the ensuing years. It seems there is something to this effect. It is very much related to normalization of deviation.

Their study shows we are not always the best judge of our own abilities. According to Dunning, the Dunning-Kruger effect is a cognitive bias that prevents people from recognizing how incompetent they are because they lack the knowledge to know that [Dunning 2014]. Unskilled personnel may suffer from the illusion of superiority, mistakenly believing that their judgement is exceptional. This is a catalyst for normalization of deviation. Conversely, highly skilled individuals are often less confident than warranted, as they know so much that can go wrong, they assume there may be something critical they do not know.

As a simple example, an operator's lack of knowledge of chemistry might prevent them from knowing adding water into acid is unsafe, but they don't know they are not competent to make that judgement. Dunning goes on to explain that the opposite may be true for skilled people. They may assume everyone knows not to add water to acid.

An organization also has the capacity to display human psychological traits in its collective behavior. Normalization of deviation is an organizational issue that is seeded at the individual human level, but is most devastating in its effect at the organizational level. Organizations and individuals can [CCPS 2012]:

- overlook high-consequence, low probability events.
- assume that risk decreases over time.
- disregard catastrophic incident warning signs (Figure 3.1)

When all three symptoms are present risk in a facility is greatly increased (shaded area in Figure 3.1), workforce at all levels.

In a good plant setting, companies analyze hazards, train personnel, update procedures for all plant conditions, manage change, maintain equipment, and support a culture of safety at the level of effort that their organizational culture inspires at the moment. When the culture is supporting a high level of effort in all these areas and others, personnel and process safety incidents are less likely to occur and are more likely to have less injurious or catastrophic effects when they do appear.

An extended period of maintaining satisfactory production levels, process yield, and product quality with no incidents or only a few minor incidents has an odd observed effect. Employees at all levels can become less attentive to deviations in both themselves and others. In addition, if the period runs long enough, skills needed for addressing upsets and deviations can start to fade. Short-term labor reductions (a win for the bottom line) that look good while the process is running under control can lead to problems in the long-term with loss of institutional knowledge, critical expertise, and a backlog of administrative tasks (such as, action items from task safety audits, PHAs, MOC closure, and updates to procedures) that did not get finished.

Figure 3.1. Organizational Traits Promoting Deviation

Another factor can be advanced process control and safety instruments systems (SIS). These created a great improvement in operations and process safety. However, for continuous processes and tightly controlled batch process units, the long periods between upsets has actually caused companies to train workers differently. Upset scenarios have to be refreshed more often. This can be accomplished in informal training in the control room, such as a tabletop emergency response drill. Since workers do not get many opportunities to see actual process deviation events, process simulators and emulators are becoming even more critical training tools. On an simulator, a worker can have the chance to respond to many process deviations in preparation for the rare real world upsets. More importantly, workers can fail in training repeatedly until they achieve success consistently. Operating circumstances may lead to increased opportunity for nonconformity when a plant has been lucky regarding process safety incidents. Where simulators are not possible, e.g. for smaller or unique operations, having what-if scenario exercises can make operators think about potential circumstances and how they would respond.

3.4.2 'It hasn't happened yet' – Excellent Past Performance Can Fool You

With the lessons of recent history and case histories in mind, how are we sometimes lulled into deviating from policies and procedures? What are ideal conditions like at your plant? When a facility experiences a long period without near misses or actual process safety events, a certain attitude can pervade within the organization.

Are We Justifying Safety instruments Deviations with Operating History?

Humans and organizations rationalize every decision they make. Timothy Feddersen, Professor of Managerial Economics and Decision Sciences at the Kellogg School of Management, summarizes the phenomenon below:

> "Rationalization means that people are constrained optimizers, and one of the constraints [in the way of choosing a preference] is that they have a psyche that requires a rationale." [Cherepanov 2009].

We must rationalize our actions as an innate part of taking action. Most often, it is hoped our rationale is thoughtful, accurate, and helpful to our decision making. On other occasions, we might be rationalizing the performance or allowance of a deviation to meet a self-justified need. A plant or process with an excellent operating record both financially and process safety-wise can, over time, create a culture of *can do* that can lead to deviations from desired processes. Interesting anecdotes about events at *cowboy-cultured* sites are common in our industry. These are sites where exception from the mainstream becomes a sought out trait among peers in the workforce and it is often rationalized by the productivity of the plant or an admirable safety history.

Operating circumstances may lead to increased opportunity for nonconformity. Some examples of situations are provided below:

- A plant startup with a unit, vessel, piping segment, etc., out of service or marginalized in some way
- A worker finding an approved procedure step is incorrect during performance of the task
- An automated or human initiated emergency shutdown causes instability

In some cases, a process may need to reasonably deviate from process parameters. Some processes exceed safe operating limit (SOL) ranges during controlled startup and controlled shutdown phases. These controlled deviations can be planned for and special cautions can be documented in the operating procedures to denote the need for enhanced awareness.

3.4.3 Overreliance On Technology

The use of new technology to try to engineer our way around system failures or to attain greater performance levels can produce undesired catastrophic outcomes. The dangers of replacing human tasks with increased technology were pointed out as early as 1983 in the landmark paper "Ironies of Automation" [Bainbridge 1983]. Change in the workplace of any sort is stressful to both individuals and the organization. New technology is change times change—it is change squared. This is because old ways must be unlearned and new ways of varying complexity must be learned in their place for individuals and teams. When faced with new technology, people sometimes question what they thought they knew and are forced to assume different behavior patterns. This can be the source of potentially disastrous normalized deviance when the technology controls hazardous processes and chemicals.

One non-industry example of new technology causing deviation is described in Greg Milner's book *Pinpoint: How GPS is Changing Technology, Culture, and Our Minds* [Milner 2016]. Milner relates how the U.S. National Park rangers in Death Valley National Park have coined the term *death by GPS* (global positioning system). It refers to park visitors who follow their GPS and then die. Death by GPS is what can happen when the GPS fails you. It doesn't fail by being wrong exactly, but more so by being too right. The GPS does such a good job of computing the most direct route from point A to point B, that it takes a driver down roads which barely exist, were used at one time and abandoned, are not suitable for the type of vehicle, or roads that require local knowledge. Using local maps, or talking to the park rangers, if available, would make you aware that taking such roads is inadvisable.

An example of one new technology that has had a positive impact on the occurrence of deviation is the use of modernized control systems and safety instrumented system (SIS) architecture. The need for human intervention is avoided and, if a deviation triggers a shutdown, the deviation can be recorded for analysis with the unit in a safe mode. A potential problem is that operating personnel could become less familiar with correct responses to deviations, or rely on the SIS to save them from deviations or shortcuts.

The past 20 year history of the industry designing and staffing for normal conditions is a problem. Workers are less prepared to deal with upsets as SIS helps systems fail less often and automatic controls generally are more reliable than ever before. Low staffing levels combined with rare upsets that require human response can create an opportunity for deviation at a time when a process may be more vulnerable to deviations. To combat this, use more and more realistic simulators and emulators (if feasible) for initial operator training and scenario-based skill refreshing sessions.

New technology has acted as a retirement incentive for some workers. Workers reported a feeling that there was a loss of control with increased automation. Some were even bypassing control systems preferring to work in manual. They felt that the machine could not do as good a job as they did.

Organizations need to anticipate how personnel will perceive the new technology. For example, we have access to more information than we ever had before, but do we interpret that information correctly? On the other hand, better accountability with reporting can be an improvement for avoiding normalization of deviation. For example, handheld delivery of electronic procedures and associated checklists means no more handwriting errors or paper copies. We can use extensive DCS data to track process performance. Upsets and their causes are more evident.

Can new technology affect plant, and possibly societal, culture? Cellular phones, and now smartphones, have been causing deviation from longstanding internal workplace rules regarding non-business reading material and personal phone calls. Also, most personal electronic devices are not manufactured to specifications allowing them to be used in hazardous locations and may be capable of producing a spark with sufficient energy to ignite flammable vapors. Many site policies restricting use of non-intrinsically safe rated electronic devices in classified hazardous areas are regularly violated by workers at all levels of the organization. The smartphone added a new twist. Supervisors used to worry about workers reading the newspaper on the job. Yes, that is distracting one from one's work, but now there are reports of texting, movie viewing, and binge-watching of television series on the job. Which technology is considered more distracting, hard-copy print books and newspapers or internet access and texting?

Is There a Limit to Automation as a Process Performance Enhancement Tactic?

In any manufacturing facility, automation tends to reduce the number of routine repetitive jobs humans previously had to perform. This would certainly seem to reduce many deviations of one type, while introducing a set of fewer but new human deviations associated with maintaining and operating the robotics or instrumented system replacing the first human layer.

Routine activities like those involving human judgement or evaluation skills that cannot be captured by automation or those that require unique manipulative competencies that robots may never achieve are examples of human-centric jobs. Human job design will adapt to the world as it always has. However, in regard to normalized deviance and job design, new opportunities will never cease to appear.

Automation has long caused a general fear in workers that they can be replaced. Warren Bennis, a leadership expert and scholar, once summed up a vision that workers who lost their jobs due to automation can relate to [Brainy 2017]:

The factory of the future will have only two employees, a man and a dog.
The man will be there to feed the dog.
The dog will be there to keep the man from touching the equipment.

3.5 HUMAN NATURE

3.5.1 Why Choose to Deviate?

The individual decision and the corporate decision to deviate are similar, but made on different scales. It does not matter if the deviation is consciously decided upon or allowed to progress unknowingly. Repeated deviations from a business management system, regulatory requirement, or any organizationally determined and approved plan is a sign to workers and outside observers that leadership values results over rules in general. In this text, we are concerned primarily with rules related to process safety, environmental protection, and personnel safety being perceived as less important than production.

John Banja of the Emory University Center for Ethics wrote a paper assessing normalized deviance in the healthcare industry [Banja 2010]. He found that the normalization of deviant work practices in an industry's internal business processes does not appear substantially different from the way corrupt practices in private business evolve and become normalized. He reports that:

"...just as the phenomena of socialization, institutionalization, and rationalization enable corrupt practices to evolve in white collar organizations those phenomena are similarly at work in the evolution of deviant behavior among (employees within the realm of their job descriptions)..."

- **Institutionalization** exposes newcomers to deviant behaviors, often performed by authority figures, and explains those behaviors as organizationally normative.
- **Socialization,** which is often mediated by a system of rewards and punishments, aims at determining whether the newcomer will or will not join the group by adopting the group's deviant behaviors.
- **Rationalization** enables system operators to convince themselves that their deviances are not only legitimate, but acceptable and perhaps even necessary.

Institutionalization, socialization, and rationalization work in a mutually reinforcing manner to dissolve anxiety among the uninitiated by representing deviant behaviors as thoroughly rational and not immoral responses to work performance challenges (Figure 3.2)

Dr. Diane Vaughan noted that the tendencies of the social structure to produce inter-group and interpersonal tensions cause organizations to seek to gain or keep scarce resources through deviant means [Vaughan 2003]. Banja's paper delineates the mechanisms by which deviations become normalized. The statements below summarize the internal dialogues that Banja finds might apply to any individual or organization. They are excerpted here with permission.

- **"The rules are stupid and inefficient!"** A very common justification for deviating.
- **"Nobody knows where that process information is located, and it isn't accurate anyway."** Knowledge is imperfect and uneven. Many critical process documents are simply not updated as often they should be, thus providing an iterative feedback loop to their not being used. Often there is more than one approved document for workers to select from, and none match. People will use this mechanism when they choose to reinvent the wheel each time they do a task rather than update the system for consistency's sake. Justifying deviation using this mindset may be the start of the phenomenon referred to as tribal knowledge. This is when the workforce collectively holds and reinforces a skewed view of how a process or system works.
- **"I haven't used this new software / equipment / controller / tool / chemical much."** The work itself, along with the introduction of new or more complex technology, can disrupt work behaviors and rule compliance.

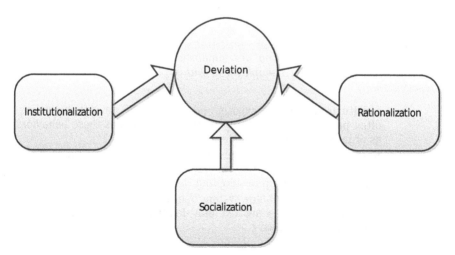

Figure 3.2. Three Phenomena that Enable Deviation

- **"I'm breaking the rule for the good of my company / shareholder / personal wants or needs!"** Applying this internal dialogue essentially demands the rule or standard needs to be broken to meet goals—production yields / short-term profit / self-imposed schedules or wants.
- **"The rules don't apply to me."** And "You can trust me." An organization does not need to purposefully hire pathological narcissists to cultivate this idea. Any human is capable of feeling superior due to education, experience, or high moral values. Their excellence is such that that they would never even think of deviating in a certain way. When they deviate using this dialogue, they are often feeling fully justified in their rationalization of acting against their internal self-image.
- **"[Silence]"externally and internally "I can't say anything about that."** Workers are often afraid to speak up when they witness a violation. If they feel they will suffer a loss in any way, people will withhold information about deviations. For example, a worker may fear being blamed by supervision even if only due to their proximity to the deviation. Some workforce cultures actively shun or harass someone who might be seen as an inside informer.
- **"Well team, fine job. But there are too many action items in your process hazard analysis report. Can you reduce the number by half so we can approve it?"** Leadership can withhold or dilute findings on system problems. This can occur in any level of management from the supervisory level to the boardroom. In his book "Flirting with Disaster," Marc Gerstein termed this organizational behavior "politics triumphing over safety" [Gerstein 2008]. The objective is to look good to upper management or to colleagues by restricting access to information or obscuring it in some way to misrepresent reality.

3.5.2 The "On My Shift..." Attitude

The attitude that one person's or one crew's approach to work tasks is superior to another's is common. At some facilities, a shift crew rivalry has been cultivated to motivate the workers. It did not always work. This attitude can lead to propagation of deviational practices to new employees who want to fit in socially with the culture of their crew. This attitude when detected is a clue that socialization of deviance may be occurring. The group is using a system of rewards and punishments to determine whether a newcomer will or will not join the group by adopting the group's deviant behaviors.

In some organizations, it is obvious that tunnel vision can occur. Pride in production, meeting the needs of record sales periods, the pressure to stay on line and not shut down a unit, a push for innovation, and corporate personnel safety goals are sometimes emphasized over process safety. Shift crews can become competitive in these areas in such a way that it acts as a set of blinders when it comes to the subtler process safety needs of everyday operations.

3.5.3 Evaluate Decision-Making Styles

The organization's desired mode of decision-making is crucial to addressing normalization of deviation. There are generally considered to be three categories of decision styles.

- Top down
- Bottom up
- Representative

Top-down decision-making sets the outcome or results an organization seeks before engineering a process to achieve those results. This type of decision-making is often seen in Theory X top-down management styles (MacGregor 1960). This is when leaders make the decisions and transmit them to other workers to implement. Acceptance of this style is often resisted by employees. When decisions are made in this style, workers might complain of unrealistic goal setting and excessive workloads. Their work autonomy is reduced thus causing more probability to deviate. In addition, workers may have the *how to do it* skills—but not the *why to do it knowledge*—that provides a purpose. This lack of understanding the purpose behind tasks can also lead to deviations that can become normalized over time.

Bottom-up decision-making seeks outcomes using the opposite approach. Rather than set goals before engineering the process to reach them, every affected level of the organization is consulted. Leadership may make the final decisions, but it is an informed decision with input from all the stakeholders. Still, some employees may not be pleased with the ultimate outcome and could harbor an urge to deviate. Most workers feel their work autonomy is enhanced by this approach. One flaw in this method is that ultimately management may reverse decisions. Especially when an action is taken with the sentiment that it is *better to ask for forgiveness than permission*. An employee acting in this mode already knows they are about to deviate to some degree.

Representative decision-making gives decision-making authority to a select team that represents all organizational parties to the decision's purpose. One or more representatives from each department is selected for the team. Each member surveys their group for input and presents the results to the group. Consensus decision-making is another term for this approach. This approach can take more time, but tends to provide the most group acceptance. There is less motivation to deviate. But how long does it take to evoke change?

4. IDENTIFYING NORMALIZED DEVIATION

From error to error, one discovers the entire truth.
Sigmund Freud

Many chemical processing industry companies can be considered high reliability organizations (HROs). An HRO is any company or organization whose business is to manage catastrophic risk on a constant basis in order to meet their business/mission process demands [van Stralen 2018, Weick 2007]. Examples are refineries, chemical plants, commercial nuclear power plants, nuclear powered ships, aerospace industries, and hospitals. Successful HROs avoid low frequency high consequence events. High reliability organizations concerned with process safety and environmental risk management benefit from establishing methods to look for and analyze burgeoning normalized deviation - "managing the unexpected." Many of the approaches offered in this Chapter to help find deviations are already parts of an existing process safety management system or personnel safety management system.

Proactively searching for errors in work processes and giving thoughtful response to findings allows an organization to see their operations in greater clarity.

4.1 FIND TRIGGER WORDS AND PHRASES

There are two types of trigger words and phrases to look for in procedures and logbooks, or other internal communication. The first type are those in an approved document like SHOULD, MUST, SHALL and MAY. These have strong connotations that certain practices or procedures need to be applied or strongly considered. These need to be used if employees need to be realigned or if the process needs to be changed.

The second type of trigger words and phrases are those you overhear or see written in a logbook that initiate concern. A few examples might be:

- We never do that step.
- That was good enough.
- We made it work.

- You don't have to pre-heat that tank for as long as they say.
- We always leave that out.
- We always wait for the high-level alarm to sound before we transfer that batch.

These statements typically represent the existence of, or a tendency toward, normalization of deviation.

4.2 USE YOUR HIRA PROCESS

A powerful tool for deviation identification is a rigorous Hazards Identification and Risk Analysis (HIRA) program. Process hazard analysis tools such as Checklists, What-If, and HAZOP, can be used during the design phase of a project as well as for existing facilities. The design phase hazard identification can establish ways to avoid deviation. The established facility analyses can find deviations that have crept into the operations. Process hazard analysis techniques provide an organized, systematic approach to achieve the following:

- Identification of the hazards associated with processes that could cause the catastrophic release of highly hazardous chemical resulting in possible explosion, fire, or toxic exposure.
- Evaluation of potential causes and the consequences of hazardous scenarios for potential detrimental impact on people, property, and the environment.
- Documentation of safeguards and the development of practical PHA recommendations for each hazard as needed if not adequately controlled.

Following-up on implementation of recommendations to mitigate or minimize the risk of the hazards is the key to making any type of PHA effective.

During a thorough PHA, a trained team with a skilled facilitator works through procedure steps and evaluates drawings and calculations to find opportunities to reduce the risks of hazards. During the review, the team may also find errors in the organization's processes.

Although it is best practice to start a PHA with fully accurate process safety information (PSI) that does not always occur. A site may have an inadequate program to keep PSI up to date, or may become accustomed to missing, inaccurate, or poorly documented PSI. In such cases, conducting the PHA provides an opportunity to find work process deviations. If in the course of a PHA the team finds itself spending excessive time fixing P&IDs, recalculating heat and material balances, or writing operation steps that were missing or incorrect, it reveals that the work processes such as management of change (MOC) or updating operating procedures are not working optimally. These meta-deviations in the work processes can allow operational deviations that may result in an incident.

Efficiency techniques that attempt to streamline PHA revalidations may drive superficial evaluation or reliance on the existing base PHA documentation. Assuming that the existing documentation is accurate and measuring change from that baseline may save time when revalidating a PHA, but may not reveal existing deficiencies in the available documentation.

Process hazard analysis helps find deviations, but the existence of the PHA work process itself offers opportunity to deviate. High performing HRO sites keep all their process safety management program administrative procedures in force, current, and compliant in their continuous implementation with effective auditing. This can help avoid deviation at the operational levels that it serves.

4.3 DETERMINE WHICH ENGINEERING ACTIVITIES REVEAL DEVIATION

One method to find deviation is to use day-to-day engineering functions as a continuous monitor. First, an organization should train its engineering group on the concept of normalized deviance as it applies in the chemical processing industry and methods for avoiding or reducing its effects. When analytically minded people are focused on a way to improve performance that allows them to openly address error, they can apply that knowledge in many aspects of the engineering group's functions.

Some of the most effective engineering activities for noticing tendencies toward normalization of deviation are listed below:

- **Management of change** – Deviations can be detected or corrected during MOC item initiation, development, hazard evaluation, and processing the of the action items through the site MOC system.
- **Process hazard analysis** – As a subject matter expert or trained leader participating in PHAs and MOC hazard evaluations, an engineer can be aware of shortcuts in the hazard evaluations or daily operations and help correct past efforts that were insufficiently rigorous. The organization should set a tolerable risk criterion for all engineers to use.
- **Process safety information development and update** – Again, this element is often overlooked in the processing industries. If engineers do not create or deliver PSI, someone, sometime, is going to spend a lot of time and money to recreate it, if they can. Engineers can choose to value timely updates as a part of their goals. It helps to engage them in this effort when the organization values timely updates too.
- **Compliance with standards and codes** – adhering to the requirements of applicable third party standards and codes can prevent deviation in process designs and modifications from creeping in.

Sometimes the engineering team is shackled by outdated engineering standards, and has a culture that is over-reliant on standards, which can result in less PSM team involvement. On occasion, engineering standards are issued with little, if any, understanding of their impact on existing equipment. This leads to grandfathering older equipment and creating disruption in the homogenous application of standards. The 2005 BP Texas City incident involved equipment that was no longer considered good engineering practice. Atmospheric blowdown stacks were still used in the BP Texas City plant for hydrocarbon venting when much safer alternatives were available and used widely in more modern installations.

The overall quality of the plant PSI in general is a revealing sign. It is essential to have accurate process knowledge to find deviations, but is a site's design basis documentation just a description of what has been installed? Always question whether a replacement in kind is the right replacement in kind. Erroneous assumptions have caused incidents when organizations accepted replacements that had not been evaluated thoroughly. Substitutions have been made without appropriate evaluation that they are true equivalents.

4.4 USE BEHAVIORAL SAFETY TECHNIQUES

Manufacturing facilities of all types are implementing behavior-based safety practices as a way to utilize behavioral analysis to effect behavior change. These techniques can assist in preventing deviations from becoming normalized through rapid intervention when they are noticed [CCPS 2012]. Behavior-based safety programs teach ways to recognize unsafe behavior and provide feedback to fellow employees whenever it is observed. Successful programs inspire participation from all workers. The core belief is that for every incident— whether process safety related or personnel safety related—there were many more unsafe behaviors that did not result in a recordable incident.

As noted by the Cambridge Center for Behavioral Studies, these programs focus on three things [CCBS 2018]:

- **Environmental changes**, that is: What is it that leads to a given behavior?
- **The behavior itself**, Am I doing my job task in accordance with the organizational procedures and values?
- **The consequences of behavior**, Were there any positive or negative responses that occurred as a result of doing that task?

When a person, a team, or an organization develops a high level of OD, that is, when all players consistently characterize the values that support the performance of all tasks correctly every time, better organizational performance is to be expected by all workers. This high rigor equates to increased accuracy in executing the work processes which in turn leads to fewer deviations and consistent product quality. Some benefits of well-established and thoroughly monitored behavior-based safety programs follow:

- Reduced injuries and modified employee behavior accomplished by reinforcing safe work practices and eliminating at-risk behavior

- Reduced costs related to injuries and incidents
- Improved communications skills among all workers
- Heightened overall safety awareness
- Increased observation skills
- Improved leadership skills
- A reinforced organizational commitment to safety

4.5 REVIEW YOUR WORK PROCESSES

If your organization has implemented a new manufacturing strategy lately, such as lean manufacturing, it almost certainly involved identifying, tallying, and examining your work processes for waste. A company can use that opportunity to increase profitability and to look for deviations in processes that, if corrected, can increase process safety performance. Those corrections, however, need to be reviewed using the MOC process to be sure the process safety implications of the changes are dealt with.

New site acquisitions or poorly managed integration efforts across multiple manufacturing sites can lead to deviation. Is there an existing corporate culture at the site that promotes process safety and personnel safety? Is the company a collection of disparate sites, each one defining its own culture and work processes? (This can occur if the company has acquired many of the sites from other companies.) All work processes (procedures) exist somewhere in the procedure lifecycle, as is depicted in Figure 4.1. This review cycle can be applied to all documents considered part of a business management system. Be alert to deviations in all stages of the document life cycle.

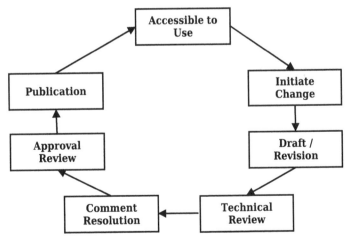

Courtesy AntiEntropics, Inc.

Figure 4.1. The Process/Procedure Life Cycle

4.5.1 Evaluate Operations Tasks

Operators have the most intimate interface with the process units. Some tasks are performed each shift or more often. Some tasks relate specifically to collecting data to reveal deviations or support process control instrumentation readings. When tasks are performed very frequently, shortcuts may be encouraged to save time on tasks that hold little interest for the performer or are perceived as low value-added work. When rare tasks are performed, especially critical ones, deviations might be made from inexperience or competency issues. Other situations that lead to short cuts include the following:

- It is possible that the procedures do not reflect operating and maintenance steps and conditions realistically.
- Operators might be faced with equipment that does not operate the way it should.
- Ineffective training is common in industry. This is a double loss in that the workers may not have the new knowledge and skills after the training session, but the cost of providing poor training is almost the same as that of effective training. Effective training should result in a positive return on investment.
- New equipment or technology can interrupt compliance with the desired work process for several reasons. Perhaps the organization did not prepare the workforce well or the vendor did not offer good client support.
- In order to get the work done, employees feel they have to bend the rules. Consider actions that the leadership would prefer they take when this phenomenon exists in the organization. What should happen instead of bending the rules?
- Sometimes senior operations staff adopt the mindset, I am experienced. I don't need to follow the procedures.
- Peer pressure can arise in a plant culture that discourages speaking out when there are problems.
- Technical experts can adopt the mindset that their knowledge allows them to bypass procedures.

4.5.2 Operating Procedures

How easy is it for an employee to request a change in procedures or training materials? How welcome is their request? Is there peer pressure to keep quiet about deviations from procedures? How often are procedures checked for accuracy? Are linkages and call-outs between different procedures correct and kept up to date? As noted previously, mismatched procedures as compared against approved process safety information indicates a failure in the MOC/PSSR elements. During walk-arounds, supervisors and managers can check if procedures are understood and being followed. Arguably, effective MOC/PSSR is both the energy source and the guiding process for maintaining a total process safety system in its optimal state of effectiveness with rare deviations.

Observation and auditing are your two primary tools in identifying deviations in procedures. Some detailed approaches are suggested in this section. Effective procedures should display the following characteristics [CCPS 2012]:

- Describe the intended activity as well as the process or equipment.
- Describe controls in sufficient detail such that employees can understand how process hazards are managed.
- Provide instructions for troubleshooting for specific scenarios when the process does not respond as expected.
- Specify when an emergency shutdown should be executed.
- Address special situations such as temporary operation with specific equipment out of service.
- Describe the consequences of critical steps not being followed and of the process' deviation from the operating limits, as well as the steps required to avoid or correct such deviations.
- Define steps required to safely start up, operate, and shut down processes, including emergency shutdown.
- Describe activities such as periodic cleaning of process equipment, preparing equipment for maintenance activities, and other routine activities.

For many facilities, operating procedures are the easiest items to check on a regular basis for deviations in their implementation. A worker can observe another employee perform the task against the procedure to evaluate whether it was followed. Although this is easily done, it requires commitment of time and personnel. Observations may have to be planned for some very critical tasks such as startups, shutdowns, and some upset conditions to focus on issues. Consider using established behavior based safety systems and associated observations as a starting point for these efforts [CCPS 2011].

The degree to which operating procedures are followed can be spot-checked using what some companies call a task safety observation (TSO). Review an operating procedure. Then, with the procedure in hand, observe an operator completing the task. After the task is complete, discuss with the operator any noted deviations. Why were the deviations made? The answers may reveal specific or systemic issues. In some industries, these TSOs or Job Performance Measures are conducted to time and evaluate the performance of key tasks that must be done in a timely manner and they become part of the regular training and requalification program.

For some critical procedures, an observer checks the task every time it is performed. Take, for example, a standard operating procedure for reactors that includes detailed operator actions for isolating a reactor prior to opening it for maintenance. The procedure task is defined as critical and requires the operator to have the procedure in hand out in the field along with the checklist. The checklist includes a signature box to confirm completion of each step of the critical procedure. In addition, a second operator with an identical procedure and checklist goes out in the field and confirms that the first operator has completed the operating procedure correctly. Both operators provide the signed checklists to the supervisor. Peer checks are very common in the pharmaceutical industry.

Of course, a critical feature of a high quality procedure is that it is current. When a PHA team or a worker finds that the procedures fail to address the current process configuration, it is clear evidence that the organization is not encouraging workers to use the procedures and the management of change process has failed. The following six steps can help you evaluate the extent of the problem due to these repeated deviations (Figure 4.2).

1. Obtain a current, accurate set of P&IDs and a comprehensive startup procedure (or any other complex process task) for a unit.
2. Mark off each piece of equipment on the P&ID as it is addressed by the procedure.
3. When finished, check the P&ID to see if the procedure failed to mention critical equipment. Was anything necessary for startup not marked?
4. Investigate each instance to determine the cause of the discrepancies.
5. Repeat this exercise with other procedures.
6. Analyze your findings to determine procedure program accuracy.

A cyclic procedure work flow process that has been proven to maintain a high level of accuracy in operating procedures, safe work practices, mechanical integrity procedures, emergency response plan procedures and all other business related work processes is shown in Figure 4.3.

4.5.3 Safe Work Practices

Safe work practices are essentially operating procedures that apply to anyone working on the site. Just like operating and maintenance procedures, the implementation of a safe work practice can be observed. Many organizations encourage behavior based safety approaches to achieve these observations. Your organization may call all combined safety related documents a safety work practice if you have adopted OHSAS 18001 / ISO 45001 *Occupational Health And Safety Management Systems* as the template for your program.

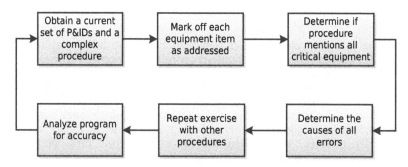

Figure 4.2. Evaluate Operating Procedure Program Accuracy

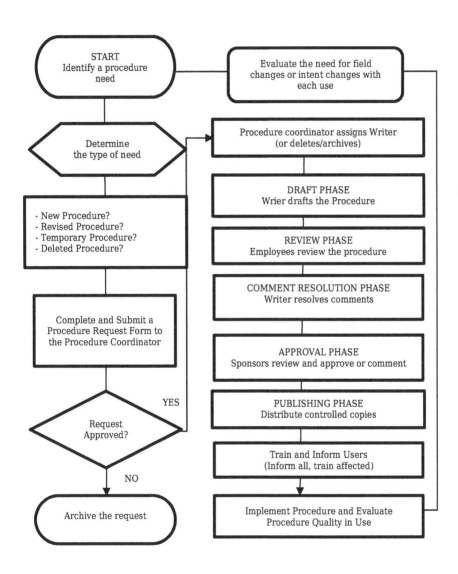

Courtesy AntiEntropics, Inc.

Figure 4.3. Procedure Cycle Work Flow Process

Some sites use periodic supervisory audits, periodic team self-audits, along with behavior based safety observations to identify deviations from consistent safe work practice application. One method that makes these more likely to be practiced is to ensure that they are written for their audience, the worker, not an auditor. Employees sometimes complain of unclear instructions in their safe work procedures.

4.5.4 Mechanical Integrity/Maintenance Program Procedures

Maintenance procedures, like operating procedures, need to clearly describe the job tasks performed by maintenance personnel in all crafts and functions of the mechanical integrity department. The general term *maintenance procedures* includes any document detailing an activity associated with the plant's current and long term mechanical integrity. It is not just about hands-on machinist, welding, pipefitting, instrumentation, or electrical tasks. Scheduling, predictive centered maintenance, and reliability centered maintenance are examples of functions with processes that can be evaluated for deviation. Typical maintenance systems in which to find deviations include:

- Documented MI management systems
- Fixed equipment MI audits
- Inspection, testing, preventive maintenance records
- Small bore piping integrity procedures
- Dead-leg identification audits
- Integrity operating window management systems
- Evidence-based inspection programs
- Corrosion under insulation inspections

One example of this approach working was when a mechanical integrity audit identified that several chemical facilities within an organization had not identified critical check valves. Inspection reports revealed that check valves were given a cursory external visual inspection, with no internal inspections being performed. A companywide initiative directed these chemical plants to identify critical check valves and perform scheduled routine internal inspections. The subsequent internal inspections revealed considerable deficiencies.

Another activity that can be reviewed for deviations is how the plant manages maintenance deficiencies. Does the plant have a bucket-list of these items that need to be fixed? That can happen, but is it growing? Categorizing inspection action items to avoid production delays can become normalized. Delaying inspections and tests can be allowed within certain windows, but it should be done properly in conjunction with the MOC process. Consider the following questions.

- Why was the inspection interval chosen?
- What should be done for similar equipment?
- Are you consistent across sites or plants?
- Why won't operations turn equipment over for testing?

Outliers become the norm when this happens. Decreased budgets can drive these deviations by rewarding staff for not doing things. Always encourage workers to understand the basis for the test, inspection, or task before delaying it. Use the site MOC process so that appropriate evaluations are conducted to ensure continued safe operation.

Mechanical integrity and maintenance procedures are work process procedures. They may include numerous subsets of procedures including testing, calibration, maintenance scheduling, pressure vessel information, critical equipment rankings, and maintenance job task procedures.

Many maintenance procedures use original equipment manufacturers (OEM) manuals to provide details. Not having them or not being available is a problem. Without OEM procedures, the industry has historically relied more upon the skill of the craftsman for day-to-day repairs and tasks rather than written maintenance task procedures. The same observation technique used for operating procedures and safe work practices can be used for maintenance task procedures where they exist. Such observations can also serve as the basis for the development of procedures that do not exist. Audits of testing and calibration programs can reveal deviations in the mechanical integrity work processes.

Paperwork documenting a repair or routine task can also be lacking. This is counter to modern predictive and reliability centered maintenance philosophies, which require detailed data from both online monitors and human interactions with equipment if they are to be applied and function properly.

4.5.5 Environmental and Quality Management System Procedures

An environmental management system (EMS) is a set of processes and practices that enable an organization to reduce its environmental impacts and increase its operating efficiency. The most widely used standard on which an EMS is based is International Organization for Standardization (ISO) 14001. Audits and observations are the best methods to find deviations.

A quality management system (QMS) is a group of work processes focused on achieving a quality policy and the specific quality objectives needed to meet customer requirements. A holistic version of quality management would include personal safety, process safety, environmental responsibility, and profitability. The ISO 9000 family of standards is probably the most widely implemented worldwide. The ISO 19011 audit regime applies to both policy and procedures, and deals with quality and sustainability and their integration. As with other procedures, audits and observations are the best methods to find deviations.

Guidelines for Integrating Management Systems and Metrics to Improve Process Safety Performance describes the synergies between performance improvement systems to help ensure safe and reliable operations, streamline procedures and cross-system auditing, and supporting regulatory and corporate compliance requirements [CCPS 2016a]. Many metrics are common to more than one area, such that a well-designed and implemented integrated management system will reduce the load on the Process Safety, SHE, Security, and Quality groups, and improve manufacturing efficiency and customer satisfaction.

4.6 USE WALKTHROUGHS AND ROUTINE INSPECTIONS

Some organizations ask employees to perform random safety and housekeeping inspections or audits. These are more like walk-throughs, or targeted walk-throughs. One tip to help with these random checks is to build a simple audit form into every safe work practice as the last attachment. In that way, an employee can print the document and have the safe work practice requirements and the audit questionnaire in hand to discuss with co-workers. Make sure everyone transmits these records to a central location for data collection after they use this tool to facilitate the information sharing as well as auditing. Methods that are more informal may also be used by supervision and management during the normal course of a business day.

Routine inspections are sometimes categorized into three levels of auditing:

- Corporate audits (generally focused on management system conformance)
- Business unit audits (focused on management systems and compliance)
- Internal site audits (focused on compliance with procedures and day-to-day implementation), which may take many forms and can include:
  o Monthly or quarterly audits by a leadership team members, staff members, operators, craftsmen, management / hourly safety committee members or a combination thereof
  o Weekly operator area audits – a housekeeping focus plus a safety equipment audit
  o Safety system audits—usually performed by the safety department and a management / hourly safety committee

Deviations can be found at all levels of auditing, however the internal site audits examine the most tasks and processes, so the probability of finding deviations here is highest.

4.7 USE YOUR PROCESS RISK AUDITS

If your organization has a process risk audit program, it regularly assesses the risks of existing and aging units and applies that learning to any new projects or major modifications it is planning. Risk audits improve process safety but, like PHAs, also can improve the operability and reliability of the plant. During these risk assessments, use the opportunity to find deviations in the way risks may have been ranked in the past. These deviations can have a dramatic impact. A process safety audit identifies alarms and interlocks that have been included as layers of protection in a PHA. A review of the process safety information, operating procedures, and distributed control system (DCS) for these layers of protection may reveal inconsistencies in the level of hazard escalation that could occur before an alarm will activate.

> The PHA of a storage tank of toxic liquid identified overfill as a relatively high risk consequence event and listed two independent layers of protection to prevent overfill. Using the scenario based defining statements to identify the probability level, the risk was deemed acceptable. The corporation changed from using qualitative risk assessment to implementing semi-quantitative and quantitative risk assessment with a robust layer of protection analysis (LOPA). A thorough risk audit using LOPA revealed that the operator was using the high-level alarm as a warning that the filling operation was complete. It also identified that the storage tank high-level interlock was out of service due to failed components. The result of this in-depth field observation risk audit and application of LOPA instead of qualitative judgement revealed that human error alone could cause a serious toxic release.

As discussed in the CCPS *Guidelines for Initiating Events and Independent Protection Layers for LOPA*, there are times when the site may need to go beyond the simple limiting rules of LOPA to include the use of Quantitative Risk Assessment (QRA) [CCPS 2015b]. QRA assists in evaluating the failure probability of a specific Independent Protection Layer (IPL) and can place the equipment failure or human error in context of its risk significance. Fault Tree or Event Tree Analysis complemented by Human Reliability Analysis (HRA) look at sequences of progression from initiating events and provide logical structures for combining the influence of different system failures, human errors and process conditions.

Using long, protracted checklists in an audit can be self-defeating. When conducting inspections, if you are looking for everything, you'll see nothing. A better approach is to use a smaller list of issues. When an audit team makes the plant inspection tour, have each team member focus on a different concern, e.g., fire protection, electrical, piping, etc. After the inspection, a list of the deficiencies can be collected for further review during the audit, or if outside the audit scope, formally passed onto plant management.

Companies should consider having some audits done by a qualified third party auditor. This provides a set of fresh observations and opinions

4.8 PAY ATTENTION TO NEAR MISSES

While a deviation is an event that may not have any discernable effect, a near miss is an event in which property loss, human injury, or operational difficulties could have plausibly resulted if conditions had been different or the event had been allowed to progress. Some examples of near misses include:

- A dropped object that does not impact personnel or process equipment
- Excursions of process parameters beyond pre-established critical control limits
- Release of less than threshold quantities of materials that are not classified as environmentally reportable
- Activation of layers of protection such as relief valves, interlocks, rupture disks, blowdown systems, halon systems, vapor release alarms, emergency shutdowns, and fixed water spray systems

Reporting and analyzing near misses provides companies with knowledge that can prevent future deviations (and near misses and incidents). Investigating near misses is a high value activity. Learning from near misses is much less expensive than learning from incidents. It is far better to have an employee report too many near misses than to not submit the key one that may have avoided an incident.

Figure 4.4 illustrates the principle that the occurrences of near misses are far more numerous than actual incidents. Thus, the opportunities to use near misses to identify deviations are more numerous than from incidents. Help all employees lose the fear they may have of reporting a near miss when they see it. Near miss reporting and analysis is critical for identifying normalized deviance.

One example of a near miss being avoided through near miss analysis is a site investigating activation of a high level switch in a tank. A finding is that the operators have routinely been using the high level alarm (90% tank level) as indication that the tank is full rather than monitoring the tank level and stopping the transfer at 85% as stated in the operating procedures.

Not all near misses need to be investigated to a root cause, but trends of normalized deviance could be investigated in aggregate for big-data trends then investigating where data suggest significant deviation. Additional near miss examples and trending metrics are provided by the CCPS [CCPS 2018d].

Figure 4.4. The "Safety Triangle" for Incidents, Events, Near Misses, and Deviations

4.9 USE YOUR INCIDENT INVESTIGATION SYSTEM

The incident investigation system is a direct means to evaluate normalized deviance at your site. It is really an incident learning and prevention process. The intent of the process is to learn from incidents in order to prevent their recurrence and to identify events and conditions that could result in a loss. Continuous improvement is supported through:

- reporting incidents
- investigating incidents to identify root and latent causes
- analyzing near miss data
- implementing corrective actions to prevent recurrence in all similar installations
- communicating lessons learned
- detecting trends through the statistical analysis and processing of incident records

One example of how investigation can be used to uncover deviations occurred when a company investigated a finger amputation. They discovered that the open hatch involved in the injury was added to the process because the quality assurance (QA) lab personnel did not want to put on the additional personnel protective equipment (PPE) needed to take the process sample at the designed sample point.

By evaluating historical records of incident investigations, an organization might be able to detect a trend that points to normalized deviance. The typical initial incident report documents information vital to the investigation. A review of the initial incident report description might highlight deviations identified by the witnesses or crew on shift. Look for trigger words that imply a deviation seemed normal to those involved. As discussed in sections 2.3.1 and 2.3.2 on the NASA Challenger and Columbia events, several instances of near misses were precursors to the eventual catastrophic failures were treated as acceptable deviations when they could have instead revealed the imminent conditions. The statistical analysis and processing of incident and near miss records to detect trends is critical to the prevention of normalized deviation and, ultimately, catastrophic events.

The final reports provide long-term data records for future analysis of normalization of deviation. Addressing the recommendations is the critical step in defending against deviations found during the investigation.

Two essential steps to prevent recurrence of an incident and any associated deviations are documenting the incident investigation findings and reviewing the results of the investigation with appropriate personnel. Ensure a system exists at your site to promptly address and resolve the incident report findings. Monitor corrective actions to help ensure completion on or before the assigned completion date. All corrective actions should be documented.

The CCPS provides extensive guidance on how to effectively investigate incidents [CCPS 2003]. Additional approaches are provided in the literature [Vaughen 2011, Klein 2016].

4.10 EVALUATE MANAGEMENT OF TEMPORARY CHANGES

A review of the types of temporary changes in your management of change (MOC) system can be an eye-opening exercise. You might find that the same temporary changes made repeatedly. Consider an engineering review of how the organization might benefit from a more permanent fix. Questions to ask include:

- Temporary repairs – how long can they be used?
- What is your criteria for permanent versus temporary MOC?

A temporary change is any non-permanent change. Special unit tests can be temporary changes even if nothing is different other than the operating procedures your team needs to use to achieve the test conditions. Other examples are new treatment chemicals or raw materials under evaluation, temporary compressors, electrical and piping jumpers, portable tanks, and temporary line repairs.

> A pipe in hazardous service had thinning below the minimum allowed thickness (T-Min). Approval by the corporate subject matter expert was granted to allow an engineered fiberglass wrap type repair for a period of 6 months. Two years later the subject matter expert was again contacted with a request to approve it for another 6 months! The temporary repair approval process was not being followed and the absence of a leak had normalized the unfulfilled need to do a permanent repair.

A temporary change needs to be approved in accordance with your management of change procedure and should have a terminating condition, an expiration date, or removal date specified. A temporary change can be made a permanent change only when addressed in accordance with your MOC procedure. Do not let temporary changes normalize into permanence without asking why we chose to deviate in the first place. Also, watch for overuse of temporary or emergency change requests, this can be a sign the normal MOC system is too onerous.

5. TECHNIQUES TO REDUCE OPERATIONAL NORMALIZATION OF DEVIANCE

The solution of every problem is another problem
Johann Wolfgang von Goethe

The methods and practices that help organizations reduce the phenomenon of normalized deviance in operations consist of many practices that are already a part of a mature process safety and environmental risk management system. This Chapter addresses selected process safety elements and how they assist in reducing the acceptance of deviations in the operations work processes along with other specific techniques you might choose to use.

5.1 REWARD RIGOR IN YOUR MANAGEMENT OF CHANGE PROCESS

MOC reviews should not be simply a paperwork exercise. Nor is it something to be done after the fact to document changes. Unfortunately, many sites slip into these modes due to normalization of deviation. Rigorous management of change should be used as a process to evaluate and manage changes to the process configuration and all of the associated procedures, training modules, and process safety information that document the organization's understanding of its processes. To improve this situation, attempt to change the paradigm of the management of change process at your site to one where your employees expect excellence. Provide sufficient resources, audit MOC performance regularly, and respond to deviations with effective action items.

A success strategy noted at one chemical plant was an MOC coordinator who was a thorough gatekeeper. If an MOC was submitted without the required level of detail he would refuse to process it. The culture slowly changed to where people would apply more rigor to the level of documentation.

Another success strategy at the same company was a mechanical integrity coordinator who would refuse to approve or extend temporary repairs unless a rational, defendable, and detailed documented engineering basis was provided.

This approach helped fend off the normalization of—*it hasn't failed yet so it must be ok to extend it again.*

Management of all varieties of change should be rigorously practiced. Whether starting up a process safety system at a new chemical processing facility or engineering or maintenance group, or managing a mature process safety and environmental risk management program, it is clear that managing changes to the total work process configuration is essential. It is also essential to evaluate every change to process design and operating tasks. The specific manufacturing process documentation and maintenance records are extremely sensitive to change. For each change, ask the following two questions.

- Will this modification to equipment, technology, or chemicals affect any worker's job tasks?
- How can we expect employees to avoid deviation in performing their tasks if the organization does not perform its task to document the preferred methods?

Refine the management of change system so that it is essentially audited every time it is used. This is a mindset to develop across the workforce— including leadership. Similarly, each time a worker uses a procedure or a drawing, they should analyze it for accuracy and ease of understanding and offer improvements if they can.

5.2 LEVERAGE YOUR NEAR MISS REPORTS

Investigating near misses is a high value added activity. One process safety expert calls them *gifts from the process safety gods.* Learning from near misses is much less expensive than learning from incidents. A near miss is broadly defined as an event in which property loss, human loss, or operational difficulties could have reasonably resulted if conditions were different or had been allowed to progress.

The reporting and analysis of near misses provides valuable information that can help prevent future near misses, incidents, and operational interruptions. The first step is reporting the near miss and then investigating to determine the causes and underlying reasons why it occurred. A thorough investigation to discover root causes helps investigators identify management system weaknesses that lead to near misses and potential incidents.

One area where near-misses are commonly not reported is during check-out and start-up. There, non-conformances are expected and are not recognized as potentially normalized deviations. Do errors in design get back to the engineer or designer so they are aware of them or are they just fixed in the field? Is equipment that was accepted at delivery as correct found to be wrong at installation? (Ordering or receiving error?) Even scheduling (assumed time to complete an activity vs reality) is a normalized behavior. "Management won't approve a schedule where control system verification takes more than a defined number of days"

Many companies recognize that there are obstacles to reporting near misses. The following is a list of the obstacles that lead to low reporting rates gleaned from surveys and from the experience of the surveyors. These keep the reporting rates low:

- Fear of disciplinary action
- Lack of management commitment and follow-through
- High level of effort to report and investigate
- Fear of embarrassment
- Lack of understanding of what is a near miss
- Organizational disincentives for reporting near misses
- Multiple incident investigation systems that are confusing

Removal of these obstacles has shown proven results. Some solutions may fit one company's culture better than others may. Improving the quality of near miss data and then gathering and analyzing it all to extrapolate the related root causes can be used to improve your processes. Categorize those you and your team identify as potentially severe near misses for future action, review, and communication. Establish criteria that might lead your organization to do a deeper incident investigation. Consider using Accident Sequence Precursor analysis as a means for understanding how these 'near misses' could have propagated into a severe incident [Collins 2016].

5.3 USE BEHAVIORAL SAFETY OBSERVATION DATA

When collecting behavior-based safety performance data, ensure the program differentiates between a near miss and a behavior-based management observation.

Many companies have implemented a behavior based safety system to have peers observe and try to correct the behavior of peers by coaching or other means. This is part of a behavior based safety management system. These observations can be assigned to the non-incident portion of the error pyramid. However, these observation data can reveal a plethora of deviations which may reveal weaknesses in process systems such as the examples provided below. Look for trends in your data to interpret the correct action plans:

- Procedures that are unclear or the interface is very user unfriendly.
- Lack of differentiation between personal and process safety. (Help management recognize that personnel injury rates do not guarantee good process safety.)
- Lack of proper equipment and effective training that could be causing the behaviors.
- Safety equipment is regularly blocked.
- Unapproved handmade tools are used in the units.

5.4 USE CREW DISCUSSION SESSIONS AND TRAINING

Actively engage the operations workforce in fighting the spread of normalized deviation. The equipment operators are the first intelligent line of detection for physical deviations that may otherwise go unnoticed. The best operators tour the area frequently to look for small differences. The operators also know when procedures are being used and when they are not. In addition, they know when not to use an approved procedure that is not up to date. Management needs to encourage the operators to identify and fix these deviations. This can be supported by a targeted training module on the topic of normalized deviation. It would be suitable for any employee in the organization.

Every shift turnover meeting is an opportunity to establish this layer of protection against normalized deviance. First, a company can train the workers on the concept of normalization of deviation and how to hold this short discussion each shift. A company can simply add the topic of normalized deviance to the standard shift turnover discussion topics. Log all deviations reported and days where none were found. A trend study of deviation types over time can be built in this manner. Some tips include:

- During all operations meetings, no matter the topic, listen for the trigger words and phrases.
- Assign crews to design special safety projects to help fight normalized deviance. Look at the topics they pick. That alone can point to normalized deviance.
- Hold facilitated discussions to audit and evaluate the concept of normalized deviance.
- Have process safety staff at all levels visit the operating areas outside of the Monday to Friday workdays. They can stay late and meet with the back shifts or stop in on Saturday to observe fieldwork. Initiate discussions related to helping avoid normalized deviations.
- Make sure management is hearing the discussions and a process exists to address them.
- Communicate the findings from equipment deficiencies reported by operating personnel to maintenance or supervision. That broadens their understanding of acceptable conditions, or helps them point out where equipment condition deviations are being normalized, "That pump's been noisy since day one."

5.5 EMPHASIZE EMPLOYEE PARTICIPATION

Engaging the entire workforce in process safety is critical to reducing normalized deviance. This is why CCPS expanded the name of this element to *Workforce Participation* in the *Guidelines for Risk Based Process Safety* book [CCPS 2007]. Your site process safety steering committee (or a similarly purposed group) is the recommended team to consult for process safety employee participation purposes. This group encourages employees to participate through recommendations, questions, or comments submitted to leadership, the process safety manager, or their direct supervision.

Membership in the committee should include representatives from the following personnel groups:

- current leadership team
- operations personnel
- reliability & maintenance
- supply chain management
- environmental health and safety

An organization should train everyone on process safety to the level of their participation. Without doing so, the organization can hold no real expectations that deviations will not occur. Without this training deviation is encouraged. This is similar to the incident command system training model. Know your role in PSM the way you know your incident role (and back up roles) in an emergency response. Train everyone on the concept of normalization of deviation and implore them to raise a flag every time they see erosion in the organization's work processes.

The following list of opportunities to engage employees in process safety and emphasize its importance, reveals how involved employees will be in a high quality process safety management program. The simplest management system, customized to fit your organization, can still be very complex and time-consuming in its implementation.

Process Safety Information (PSI). The process safety information has to be compiled, retained, and made accessible for use by the organization and contract employees. Guarantee continuous employee access to the PSI. Using operating personnel to regularly review and report errors and omissions in PSI can be an effective way to engage personnel and maintain a higher level of accuracy on PIDs, Encourage them to request changes when they find they are needed.

Process Hazard Analysis (PHA). Employees with appropriate expertise in the process being evaluated are selected to participate as members of the process hazard analysis (PHA) teams. New employees on a PHA team should be encouraged to ask why steps are done a certain way, how equipment can fail, etc. Employees selected to participate can include any salaried or hourly personnel. Employees can be directly involved with the implementation of the PHA recommendations.

> A facility invited both new (less than two years' experience) and experienced operators to a HAZOP of an ammonia refrigeration system. At the end of the HAZOP the new operators said they learned more about the system in one week than they had in two years.

Operating Procedures. Employees who work in or maintain hazardous processes must have available written, accurate, and up-to-date operating and maintenance procedures that correspond with their work activities. After training is complete, employees are expected to operate equipment according to these procedures. Operating, maintenance, and technical staff personnel participate as subject matter experts, procedure reviewers, and writers to develop, validate, and update the procedures. Any employee may request a change to a procedure.

Training. New employees receive orientation training and process overview training for the areas to which they are assigned. Employees assigned to a work in a hazardous process will receive the appropriate level of training before being involved in operating a hazardous process. Initial and refresher training can be implemented and documented. Encourage employees to provide feedback after selected formal training sessions to evaluate and improve their training. They are also consulted with on the frequency of refresher training.

Contractors. Employees have an integral role in the interface with contractors working at the site. Your employees often work side-by-side with contract employees. This provides them with a high level of knowledge about how the contractor's work can affect or be affected by the process involved (or adjacent processes). Each site employee and contract employee is encouraged to have a questioning attitude and to escalate concerns about unsafe work activities or conditions.

Pre-startup Safety Review. Employees with appropriate knowledge and skills participate in completing pre-startup safety reviews (PSSR). Gaps found during PSSRs may indicate normalization of deviance by any part of the system that delivered the new or modified equipment. Begin each PSSR meeting reminding the team they each have the right to delay startup until there is evidence that all hazards and requirements are adequately addressed, regardless of schedule pressures. Encourage employees to have a questioning attitude, do informal pre startup safety reviews of their own, and initiate a dialogue with cognizant personnel on any concerns identified.

Mechanical Integrity/Asset Integrity. Employees are involved in the daily maintenance of hazardous processes. Involvement includes employees from groups whose knowledge and expertise is required. Affected employees will receive training in the various elements of the mechanical integrity/asset integrity element.

Safe Work Practices/Hot Work Permits. Employees and contractors will be trained in the use and requirements of safe work practices including hot work permits. Employees may direct any questions concerning safe work practices to management or directly to the safety department personnel. The written procedures for safe work practices are available electronically on the organization computer network and in hard copy.

Management of Change. Organization employees consult on the development and revision of the management of change process through the PSM steering committee. They may initiate the management of change procedure using the management of change (MOC) form. Communication with employees prior to implementation is a critical element of MOC. Points to consider when informing and training employees during changes include:

- **When To Inform Employees About A Change** - Employees whose jobs may be affected by a change in process safety information resulting from a change must be informed of the change.
- **When To Train Employees On A Change** - When of the following items are developed or revised due to a change and that change affects an employee's job, the employees need to be trained on the change prior to working to the modified process (check for employees who have been on long weekends or vacations):
  - o Process overviews
  - o Process safety information
  - o Operating procedures
  - o Safe work practices
  - o Maintenance procedures
  - o Emergency response procedures

Incident Investigation. Members of the joint health and safety committee and selected employees with appropriate knowledge and expertise may participate as members of the investigation team and provide subsequent follow up for plant incidents. Incident investigation reports are reviewed with personnel whose job tasks are affected by the findings. Employees may access information about incident investigations on the organization computer network. Deviations tied to incidents can be isolated and addressed.

Emergency Planning and Response. The emergency response plan is available electronically on the organization computer network and in hard copy for review by employees. It may be written using an Integrated Contingency Plan (ICP) (an example ICP format has been developed by the U.S. National Response Team to consolidate U.S. regulatory requirements related to emergency response [NRT 1996]). Employees will be trained on the specifics of the emergency response plan applicable to their work areas. Employees participation in all training exercises and evaluations is needed.

Compliance Audits. Compliance audits are intended to verify that the procedures and practices developed under the standard are adequate and being followed. Employees trained in the requirements of the process safety management program may participate as audit team members. Others may be interviewed as part of the audit process. Employees who may be interviewed should be made aware of the need for their honest and frank input to the auditors' questions. Employees can be consulted on development of this element through the PSM steering committee. Employees should be trained in auditing techniques before participating as audit team members.

5.6 ENCOURAGE OPEN DIALOGUE SUPPORTING ALL WORKERS WHO RAISE NORMALIZATION OF DEVIATION ISSUES

In order for an open dialogue to be effective, management needs to be open and willing to address the deviations identified. There should be clear expectations for all roles and responsibilities. This requires that the leadership teams show in their actions that they value employees who proactively seek to reduce deviations and, most importantly, improve systems to reduce the opportunity for error.

One effective tool to encourage dialogue is a safety blitz. On a periodic basis a group of volunteers goes all through the unit looking for anything that is not right. It could be grating clips missing, poor lighting, fall protection conditions, signage, eyewash stations, and safety showers, motor control center (MCC) doors propped open, condition of tools—anything. All observations are documented, shared with the workers, and everything is addressed.

Another good tool is a job cycle check on operating procedures (or any work process procedure). Hold open discussions on how work is performed and whether there are shift-to-shift differences. It is critical for management to understand and discuss openly how to best operate the plant and get worker input. As before, it is human nature for people to resist when being told what to do. Not everyone is going to get their way necessarily, but they may come to respect the process.

5.7 LEVERAGE LEARNING FROM YOUR PHA PROCESS

Process hazard analyses (PHA) are excellent learning opportunities for all members involved as well as the organization. A team consisting of process experts, experienced operators, process safety professionals, and other subject matter experts is assembled and guided by a PHA facilitator to rigorously evaluate the process safety information, identify initiating failures, ultimate consequences, and a comprehensive list of safeguards and controls. This detailed analysis is a perfect opportunity to review operating procedures in detail and listen for trigger words that may indicate operators are not following the procedures as written.

Anyone who participated in the rigorous PHA method such as a hazards and operability (HAZOP) study at a chemical processing facility in the 1990's may have noticed a difference in how they are conducted today as compared to how they were performed then. One difference is in organizations' reduced interest in the operability aspect in favor of the hazards aspect of the study. Studying the operability aspects in a PHA reduces the chance of finding where deviance has become normalized. The list below shows some deviations that the PHA can reveal, as well as deviations in the PHA implementation process itself.

- Missing operating instructions
- Out of date operating instructions
- Out of date process safety information such as P&IDs
- Poorly documented incidents and near misses
- Failure to communicate previous incidents to those who need to know
- Missing or incomplete kinetic and/or thermal stability data
- Abandoned in place lines or equipment that may actually be used
- Discovery of unmanaged changes that have been made in the system
- Lack of management commitment in conducting and acting on PHA recommendations
- Inadequate training or understanding of new technology
- Inadequate PHAs due to:
  o irregular PHA meeting schedules
  o wrong PHA procedure chosen
  o inappropriate team experience
  o knowledgeable process person or key people unavailable
  o scope too restrictive
  o inexperienced leader
  o sessions too long
  o some identified hazards dismissed as being unimportant
  o poor documentation
  o participants quashing other's opinion
  o failure to fully consider the effects of cross contamination, contamination by service fluids or contamination prior to or during the receiving of raw materials
  o failure to consider persistent operability problems (these can sometimes result hazard scenarios)

5.8 PERFORM A JOB TASK ANALYSIS FOR EVERY JOB POSITION

As with any industrial or technical training analysis, it is often easy for management to simply list the things that it assumes a worker needs to know to do their job safely and efficiently. This is usually off base by a large percentage when measured by the knowledge, skills, and attitudes (KSAs) a worker needs to be fully engaged and confident in his or her role in their job. A Job Task Analysis provides a better view of the KSAs for each position.

The chemical industry in general has an advantage; it requires extensive procedures for work tasks at an appropriate level of detail. These procedures become the heart of any training system. When it comes to process safety training, it is valuable to start by providing new employees with a general overview of the specific hazards at a facility Existing employees receive refresher training on a regular basis. More valuable yet is encouraging workers to train on and use well-written procedures that warn of conditions that might put workers or the public in harm's way from releases, fires, explosions, or exposures. Effective training on procedures with performance evaluation is key to success in developing a safe and proficient workforce in all positions, from the laboratory to commercial production. Prioritize the types of positions to address first and use the data to create or update competency logs for each position to accompany the employee curricula.

The job task inventory and competency log provide a baseline against which deviation can be measured. It speaks to this problem: an organization cannot address deviations if you do not first identify the desired norm. Determine which categories of tasks are most critical and start analyzing those first. See the basic organization of the job task analysis data in Figure 5.1. A more detailed description of the steps for performing a job task analysis is provided in Appendix B - *Job And Task Analysis*.

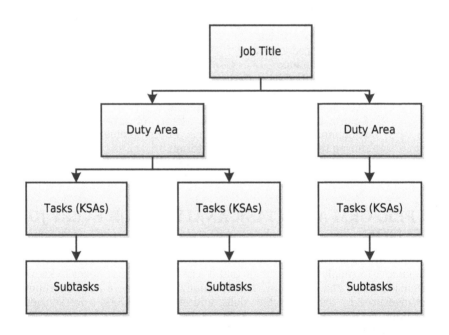

Courtesy AntiEntropics, Inc.

Figure 5.1. Job Position Tasks, Subtasks, and KSAs

5.9 RECOGNIZE ALL WHO COMBAT NORMALIZATION OF DEVIATION

If your site establishes a normalization of deviation monitoring effort, make sure it includes effective and meaningful recognition for any employee who makes significant contributions to identifying or correcting deviations. The measure should be whether they succeed in returning work processes to their desired configuration, or initiate the improvement of a process by identifying changes that were needed. An inadequate response to a deviation can be just as bad as no response to a deviation.

5.9.1 Drive Expectations of Excellence in Routine Tasks

Recognize the day-to-day work of all employees expected to rigorously follow procedures—operators, maintenance technicians, engineers, management, support staff and others.

An example might be the need for the mechanical integrity group to keep inspections and testing up to date and all repairs addressed within the work order system. An industry survey found that most releases came from piping and storage tanks. These components are generally considered to be common and not the highest technology equipment in a chemical processing plant. The causes of failure found were just as common [AFPM 2016]:

- **Fixed Equipment Mechanical Integrity Failures** – Internal and external corrosion, erosion, and cracking were causes. Inspection programs were less than adequate.
- **Equipment Reliability Issues** – Premature failure was the cause. Maintenance or repair was less than adequate.
- **Human Factors and Performance Issues** – Valves being left open, open-ended lines, loading and unloading events, and tank filling errors were the causes.
- **Design** – Winterization procedures and specifications were not adequate.

Ask mechanical integrity inspectors to identify and report equipment that is not made available to the inspection group for scheduled internal inspections. They can also identify equipment that is not adequately cleaned to allow thorough inspections. Both situations may indicate normalization of deviation.

5.9.2 Avoid Reinforcing Behaviors that Lead to Unmanaged Deviation

As mentioned earlier in the book, avoid rewarding the shortcut taker who may have saved the day in terms of preventing an upset or meeting stretch goals for increasing profit. That employee also may have been responsible for propagating deviations that became normalized, thus creating the need for saving the day. The ends do not justify the means in these cases. Try not to punish honesty or reward liars when the behaviors are observed and assessed.

It is well known that performance measures the organization establishes drive workforce behaviors. Are your measures driving undesirable behaviors? Site leaders need to be mindful of which behaviors they reward and reinforce.

5.9.3 Check Your Metrics

Are You Measuring What You Think You Are? Are You Acting on the Results? Are they driving the correct behaviors and outcomes?

When establishing practices to curtail normalized deviance in its earliest stages, be selective in what you choose to measure. Before choosing metrics, ensure they are standardized across a company and are clearly defined. For example, to have a metric for on-time preventive maintenance of safety critical controls, you need an agreement on what the terms safety critical and on-time means. (Is it the exact day in a database or is it on time so long as it is scheduled during the week or month?) For some preventive maintenance activities, such as in the instrumentation craft, how the work is performed also needs to be standardized. Metrics are ineffective management tools if they are not reported consistently by everyone.

Another example is with a management of change metric for percent of MOC item change requests filled out 100 percent correctly. If different management of change forms and procedures and workflows for approvals exist across a company, that metric may not provide a clear picture of performance.

The same is true with tracking overdue action items. Depending on the target date for the action item, it could be late because it was too aggressively set by one manager versus another. What about the case where the target date is allowed to be moved out to a future date? Is that okay once? Twice? Never?

Managers also use metrics to benchmark their performance to similar companies in an industry segment. While this may seem routine for occupational statistics like recordable injury and lost workday rates, benchmarking process safety metrics may be less reliable due to the points above. Some possible metrics for normalized deviation detection might be the number or percentage of:

- deviations found in routine audits of the physical plant
- operating procedures or maintenance procedures overdue for periodic review
- preventive maintenance, calibration or inspection tasks overdue
- recurring incipient or degraded failures with the same failure modes in maintenance and repair records
- alarms per day
- standing alarms
- inhibited or bypassed alarms
- activations of pressure safety valves or rupture disks
- emergency shutdowns by interlock or operator intervention
- fires

Additional guidance is available to help you select leading and lagging process safety indicators based on the process risks in your organization [CCPS 2018d].

6. TECHNIQUES TO REDUCE ORGANIZATIONAL NORMALIZATION OF DEVIANCE

The trouble with organizing a thing is that pretty soon folks get to paying more attention to the organization than to what they're organized for.
Laura Ingalls Wilder

Normalization of deviation in an organization requires the collective psychological tone of the entire organization to guide it to a frequently subconscious consensus decision to accept deviations. In high reliability organizations (HRO), that condition should be curtailed at its earliest detection. The organization can do this by openly committing to reducing the effects of operational deviations as rule number one in order to prevent normalized deviance from seeping into the organizational culture.

Another characteristic of organizational deviance is that it also promotes individual deviance. Thomas Lawrence and Sandra Robinson of the Sauder School of Business, University of British Columbia, found the following [Lawrence 2007]:

"Although organizational control and power are often designed to diminish workplace deviance, they also have the capacity to incite it. This is because enactments of power that confront organizational members in their daily work lives can create frustration that is expressed in acts of deviance."

Organizational deviations that set the tone for deviation as normal are often administrative. Some examples are:

- To meet a schedule, a new or revamped plant enters startup phase before all of the process safety information is in place. This deviation from a site's PSI program procedure means the unit startup procedures (and others) are likely incomplete and ineffective.
- Inspections, preventive maintenance items, calibrations, PHA re-evaluations or recommendations and action items (for example,

from PHAs, incident investigations, or other administered organizational processes.) are consistently late. Publishing key performance indicators (KPI) on these items can drive the group toward improvement but it also highlights and may encourage the normalization of deviations if the numbers are high and allowed to remain so.

- It becomes normal to see out of date P&IDs and operating procedures due to lack of time and human resources allocated for required update. A site's most effective and called upon operations and technical staff workers are usually the best persons to help with these tasks as subject matter experts, writers, and reviewers. However, the organization's messaged tone is that these are low value tasks.

- Important organizational management duties such as personnel performance appraisals and leadership inspections are not completed on time.

- A project is completed but certain aspects are found to be outside the company's design specifications. The engineering team, with executive management approval, decides to accept the deviations because there is no additional money in the project budget.

These organizational behaviors should be avoided as they can cultivate a culture of low operational discipline. Such a culture is one where commonly expected values and attitudes that feed success are not consistently displayed in organizational behaviors. Organizational behaviors can affect team and individual behaviors, feeding a spiral of deviation toward lower and lower OD. This is the base instinct that spawned the *"do as I say, not as I do"* idiom revealing the stereotype of hypocrisy in authority. Organization members are always subconsciously expecting high levels of OD. Its absence is instantly detectable. One example of this is a company that accepted turnover of a large capital project without receiving the entire process safety information package. It is all too common that projects are started up without thorough, accurate, as-built information in place. Capital projects can be executed without the needs of operations in mind. What should be turned over as a standard of turnkey acceptance? Operating procedures and other work processes are often not part of engineering design efforts and can be overlooked until the initial startup date draws near.

6.1 TROUBLESHOOTING

Troubleshooting can be a source of new deviation even while a team is actively attempting to correct an operational issue. Some issues, whether from a process safety stance or from the organizational culture stance, can cause:

- high anxiety
- high stress
- duress
- initiation of speed-centric actions

Those times are when humans need time to think and use approved preplanned work processes for troubleshooting. On an individual basis, deviation can be caused by solo, unmonitored actions that allow variance. When a group attempts to troubleshoot informally, a herd mentality can encourage acceptance of an inappropriate solution without full assessment of the situation and low member participation levels. Engineering work during unit commissioning is the best example of this. Controlled startup and shutdown of continuous processes are also an example. Consider using the following techniques:

- Establish a boundary between the system(s) undergoing troubleshooting and the rest of the process
- Develop operating criteria that can be put in place during troubleshooting (for example, when to initiate a shutdown or emergency shutdown).
- Establish trouble shooting guides and boundaries on how many attempts or how far the team can go before calling a time-out.

6.2 CONSISTENTLY ANTICIPATE THE HUMAN TENDENCY TOWARD NORMALIZATION OF DEVIATION

There are several agreed upon aspects of the human tendency toward normalization of deviation. Part of the organization's system to reduce normalized deviance should be to look for these attitudes in the workforce behaviors and then identify and correct the root causes of these attitudes. Some typical rationalizations for deviations are offered below.

The rules are wrong. Deviators often interpret rule compliance as irrational and a drag on productivity or efficient job performance [Vaughan 1999, Vaughan 2004, Gerstein 2008].

The rule is unknown or misunderstood. The deviator might not know that a particular rule or standard exists. They may even have been taught a process deviation without realizing that it was a deviation.

The rules changed with new technology. The deviator's knowledge is disrupted by new tools and equipment, software and interfaces. As Richard Cook states, "*When new technologies are used to eliminate well-understood system failures or to gain high precision performance, they often introduce new pathways to large scale, catastrophic failures. Not uncommonly, these new, rare catastrophes have even greater impact than those eliminated by the new technology* [Cook 2000]."

The rules do not help meet my / our goal. This is when the deviator thinks the rule or standard must be broken to progress. It is not proactively meeting the current goal(s).

The rules are for others. Deviators perceive themselves as good and decent people, they are able to think of their rule violations as rational and ethical. They see their acts as righteous and may be enraged if confronted with contrary evidence.

The rules stifle my creativity. Deviators think their professional judgment is better than the rules.

Rule breaking is ignored. Deviators are allowed to deviate by their peers. A 2005 study in the healthcare delivery system found a common theme when a person witnessed another worker deviating was that *"it was between difficult and impossible to confront people."* (Gerstein 2008). This stops the flow of information to any system that management may have put in place to find normalized deviance.

Rule breaking information is diluted. Underreporting near misses or even occupational safety incidents can happen when workers feel it will draw undue attention to them or cause a team to lose status. Sometimes reporting an incident can mean losing a safety related monetary bonus. System issues are often diluted or misconstrued as that information climbs the chain of command.

6.3 ADDRESS THE SYSTEMIC ISSUES WITHIN THE ORGANIZATION

The management system is the organization's responsibility. Its executives, directors, and managers need to set the process safety vision in their priorities. Strong leaders know that displaying expectations of excellence, operational discipline (OD) in their behaviors, and adherence to standards is their responsibility [CCPS 2011, Klein 2017]. Employees adapt to mimic their leaders.

6.3.1 Organizational Culture Shifting

If you find that an organization has a culture of low OD, that is, a culture that leads the organization to take actions that do not display communal values and attitudes to do the right things—it urgently needs to be modified. Each of us seeks OD in persons, teams and organizations. Culture change is one of the most difficult goals you can attempt. The larger the organization, the more difficult it becomes. Is your organization set up to facilitate sharing best practices? Is the company identity clearly supported by management? Has there been a flurry of retirements at a site? Have there been acquisitions of new facilities? To begin the journey of culture change, consider focusing on developing excellent accurate work processes for all critical tasks in the organization and displaying value for adhering to them in corporate behavior.

6.3.2 Plan Your Resource Allocation

As described previously in Chapter 2, much of what your organization does in regard to compliance with a basic process safety system works to help find and reduce deviations.

The process safety management systems of a complex global company as well as a small single-site company both need to fulfill the same location-specific regulations. However, the organizations may be very different in their management of resources. The most important aspect of using process safety techniques to fight normalized deviance is that the management system in place is both consistently used and updated as needed. Consider your organization's culture and resources. Your management systems should all consider these dimensions. Many companies have adopted management systems with lofty aspirations that their site culture or their company resources simply could not implement.

When planning to reduce normalization of deviation, consider the following questions:

- Have productivity initiatives been evaluated to determine if they are sending the wrong message?
- Have the proposed initiatives been discussed with those who will be accountable for implementation? Listen for hidden concerns and review the need to implement well includes safely and sustainably.
- How well do we maintain our focus and support for day-to-day manufacturing versus capital work?
- Who pays for this? How are resources allocated?
- How lean is too lean?
- Are low probability/high consequence hazards being marginalized?

Operator training is one area affected by resource availability.

- Is time given to create and implement detailed training programs or does the individual shadow someone for a period?
- Is understanding tested consistently?
- If multiple trainers are used, is there any process to audit the quality of the training? Everyone has had good teachers and bad teachers.

One thing seen in the chemical processing industry is that process safety staff position duties are often looked at as part of process engineer or safety professional on-the-job training. Large companies with hundreds of locations often do not have the adequate numbers of process safety mentors. That makes it difficult to encourage consistency.

> An example of capital resource allocation being affected by deviation involved one company auditing and discovering that the fixed fire protection system maintenance and testing program was weak. Upon identifying a program management system failure, the plant required a separate thorough review by a subject matter expert on fire protection systems. This subject matter expert review identified that underground fire mains had weak sections which were prone to rupture, some fire pumps were inoperable, and all were unreliable. Once defects were identified and communicated, funding was made available for multi-year prioritized capital project improvements. Normalized deviance was thwarted.

6.3.3 Managing Organizational Change

Managing organizational change has been found to be an extremely critical but long overlooked facet of implementing rigorous process safety. Humans are at the root of every deviation. That fact alone is reason enough to manage their turnover and workforce expansion or contraction.

Here are some questions to answer when considering organizational change impact, with "yes" answers to the following trigger questions initiating an MOC:
o   Will the proposed change affect occupational and process safety responsibilities?
o   Will the organizational charts need to be changed regarding positions that affect safety performance?
o   Will the proposed change modify, add, remove, or combine any position so that the change affects safety performance?

Evaluate the change for process safety:
- Has the change been evaluated for its general impact on safety performance and found to be of acceptable risk?
- Have all action items identified in the safety evaluation above been addressed and resolved?
- Are safety critical roles and responsibilities for the job identified? (Read *Guidelines for Defining Process Safety Competency Requirement* for a how to manual and spreadsheet tool for doing this [CCPS 2015a])
- Does the new employee have the skills and training required to fulfill the safety critical roles and responsibilities?
- Have the safety critical roles and responsibilities of the new role been communicated to the employee?
- If a position is changed or eliminated, is adequate time allowed for the individual to learn the added job tasks?
- If a position is changed or eliminated, have safety critical roles and responsibilities been assigned?
- Have documents and organizational charts been reviewed and appropriate action items initiated?
- Has a training plan been developed?

One example of deviation due to poor organizational change management involved a safety engineer who had responsibility for testing and maintenance of fire protection equipment. He managed the scheduling using a combination of personal memory and Excel® spreadsheets. He retired and his position was not filled for several months. The safety manager who assumed the safety engineer's role was unaware of the schedule for fire protection maintenance and testing. Many tasks became overdue and were eventually uncovered in an audit. This is another example of deviation being abated through thorough auditing. However, a window of unrecognized increased risk exposure did exist for a period in the plant before being uncovered in the audit process.

See the *Guidelines for Managing Process Safety Risks During Organization Change* for a description of an Organizational Change Management process [CCPS 2013].

6.4 WHEN TO STAND DOWN - HALTING OPERATIONS TO FIGHT DEVIATION

As described in the CCPS book *Guidelines for Risk Based Process Safety*, the first line of defense against catastrophe is a workforce culture that takes a questioning attitude [CCPS 2007]. A general stop work program described with the safe work practices (SWP) is one way to encourage this attitude. The document needs to empower all workers to intervene when the procedures they are familiar with and the work practices that apply are not properly followed or when they believe that the work presents unacceptable risk. This needs to be designed to foster an environment of learning and improving rather than an environment of snitching. It is vital that every employee feel that there is no blame for being corrected nor is there ostracism for identifying the behaviors called out. One chemical company defines this in their core safety principles as *you are obligated to stop a job or refuse to perform a job if it is not safe or cannot be performed safely*. The company's safety principles are stated on a pocket size, laminated card and distributed to their employees and contractors. This approach requires a workforce that is capable of identifying potential hazards that may be associated with their job's work activities, and feels free to help all employees and the organization improve safety in the long term. Always support workers that make the right choices. If they shut down a job or a unit under the safe work practice, there should be no repercussions. The organization must support the employee's authority to shut down a unit.

In the process safety field, there has to be a line that is not to be crossed. Organizations can use the stand down technique if that line is threatened or crossed. In the military the term stand down means—*a period of relaxation after one of intense alert*. It applies to battles as well as safety. After a safety incident, the military organization will take time to reduce operations for a period and arrange schedules and sessions so that everyone in the organization can be briefed on the incident to raise group awareness of a specific hazard or class of hazards.

At chemical processing facilities, if the occurrence of a certain type of near-miss is critical enough, or if a process safety incident occurs, with or without human harm, some organizations have adopted and adapted the military safety and security stand-down approach to emphasize incident lessons learned and engage the workforce in self-directed awareness raising on hazards of special importance. It can be used at one site or at every site in a company. Some examples are given below:

- A loss prevention inspector noticed unacceptable dust levels and a decision was made by management to shut down the plant until all floors were 100% cleaned up and the aspiration system was repaired.

- Operational errors were being made and costing the company a lot of money with off grade product and safety performance. The plant was shut down until everyone went through some retraining and testing.
- Housekeeping in a control room was unacceptable and the plant manager shut the department down. The employees were instructed to clean it up so that the plant could then be restarted.

Consider whether the following approach would work for your organization. Some companies use brief regularly scheduled calls where occupational and process safety incidents and near misses are discussed with senior executives. Literally hundreds of employees join these calls each time in a mini-stand down of sorts to communicate the incident and corrective action with safety professionals, engineers, and plant management. For additional information, the U.S. OSHA website has helpful recommendations for organizing a safety stand-down [OSHA 2018].

Three examples of a company addressing stand-down level deviations are provided below.

- A non-return valve has a design where a stem is able to come loose from the flapper valve, blowout through the valve casing, and release ammonia. The company has a review done at all its chemical plants to determine if any more of these types of valves are in service.
- During a power interruption, a chemical release occurs due to a control valve not activating in the fail-safe position (programming error). A company-wide program is put in place to validate that fail-safe valves will actually fail in the correct position.
- A pump in hot water service has an expansion joint on the suction line. The start / stop switch is located adjacent to the pump. When the pump was being started, the expansion joint failed and sprayed hot water on the operator. A company-wide program was put in place to eliminate unnecessary expansion joints, to relocate start / stop switches away from the direct line of fire of the equipment, and to implement a routine replacement program for expansion joints.

6.4.1 Include and Enforce 'Do Not Exceed' Points in Critical Procedures

Critical procedures, for example startups, shutdowns, critical compressor operations, and some maintenance tasks, all need very clear *do not exceed* warnings in the body of the procedures whenever there is a known risk of realizing a hazard by violating design parameters. Such warnings might be given for DCS permissives within the safe upper/lower limits that can be used to define operational boundaries. Some companies password protect critical parameters. Property insurers have recommended having signs in the control room stating that the operators have the authority to shutdown rotating equipment (generators, compressors, and other equipment) during abnormal operating conditions. According to some insurance carriers, there have been situations where an operator is phoning someone to ask for permission to shut down when catastrophic failures occur. This happened during the 1988 Piper Alpha oil platform explosions and fires and directly contributed to the extent of the disaster [CCPS 2008b].

6.5 PROMOTE TRANSPARENCY AND ACCOUNTABILITY

After a catastrophic incident, the incident investigators from the organization and outside agencies as appropriate immediately begin data and evidence collection. The most perishable evidence types are eyewitness account interviews and post incident interviews with other employees. Time after time, investigators find that employees whose jobs included process safety critical tasks (or tasks that supported process safety critical tasks) did not understand their role in the larger process safety management scheme. Organizations need to train everyone about the interwoven nature of the process safety elements they act within and how that meshes into meeting other business drivers such as product quality, environmental performance, and increasing shareholder value.

Transparency in how the process safety management system is built and used creates a desire in the workforce to support it. Transparency is important in process safety because the protective layers are not as obvious as they are in occupational safety. Guards and physical warning signs are visible. DCS and SIS software are not. Operators do not know if process control loops and SIS equipment have had their normal preventive maintenance, or how that affects the protective layers. Not being able to picture yourself as crucial to preventing incidents can reduce your sense of accountability.

The best management systems are kept as simple as possible and are designed to meet the culture and resource levels of the organization. Many companies originally adopted approaches that were quite high-minded in what the program committed them to do to comply with local process safety regulations. Then they found that they could not possibly implement the management system as it was built. They had to apply lean manufacturing techniques to process safety compliance and then commit the resources needed. An organization's workers are the first to know when leadership states unrealistic goals or implements unworkable systems. One known building block of success in achieving transparency and accountability comes from nuclear naval and facility practices of relying on a known level of technical competency for each worker position.

6.6 ADHERE TO GOOD ENGINEERING PRACTICES

Compliance with standards is one of the twenty elements of Risk Based Process Safety [CCPS 2007]. The chemical processing and refining industries exist because of our ability to engineer creative approaches to manufacturing needed materials. In order to forestall normalized deviance, start using well vetted standards, codes, and practices for your specific technological engineering field. The design sets the foundation. If the design has flaws, operations and/or maintenance may implement work-arounds for the life of the process, which may lead to normalization of deviance or an incident. The deign therefore sets the foundation for creating and maintaining a physical and management process to help ensure you can maintain safe and efficient production for the long-term.

Using recognized and generally accepted good engineering practices (RAGAGEP) for design, construction, and management of change helps avoid such deviance. As regulatory examples, the U.S. OSHA and U.S. EPA refer to the need for RAGAGEP in these specific sections:

- OSHA 29 CFR 1910.119 (D)(3)(ii): "The employer shall document that equipment complies with recognized and generally accepted good engineering practices." [OSHA 1992]
- EPA 40 CFR 68.73 (d)(2): "Inspection and testing procedures shall follow recognized and generally accepted good engineering practices" [EPA 2017]

RAGAGEP consists of the standards and practices that your organization has developed based on industry good engineering practices for maintenance, testing, and inspection of similar processes or equipment designs. It should be used as guidance in operating and maintaining your facilities. Adhere to both internal standards and industry codes (for example, API, NFPA, NBIC, Chlorine Institute pamphlets, ANSI, and ASME). Three examples of deviations that could have been avoided by using RAGAGEP are presented below:

- A company developed minimum design requirements to upgrade the design of selected high hazard processes to bring them into alignment with industry consensus standards. The minimum design standards were developed but the implementation stagnated.
- A risk based inspection (RBI) program was implemented at a natural gas processing plant. The program implementation was weak and used as a means to extend inspection frequencies rather than manage the hazards according to risk.
- Fire protection water spray monitors were installed to be in compliance with RAGAGEP but were so close to equipment that, in the event of a fire, the thermal radiation would prevent access.

6.7 ENCOURAGE MANAGEMENT TO USE TECHNICAL EXPERTISE

Experience within the technical knowledge base of the site personnel and total organizational community should be used to its maximum effect. Follow the advice of your experts. Technical experts also need to validate their plans, perhaps because they tend to address the most complex issues. When two or more experts sit down to review an initiative or address a problem, there are two things can happen; either they find a serious flaw and revise the plan to avoid it, or they agree that the plan is complete and can implement with confidence. There's no downside to doing the review. Evaluate leadership level decisions for both the message sent to employees and the technical basis and risk level impact they might have on operations and the organization.

There have been instances when sites have been driven to push the production design limits of a unit to meet a client commitment, or more tellingly, increased output beyond the limits to take advantage of a temporary market condition that rewarded increased production. This can be hard on equipment and personnel. Normalized deviation begins with humans taking shortcuts and not following the work processes or misunderstanding of the requirement. The human element tends to fail in both of these when fatigued, as would the metal used in a reactor when weakened by repeatedly applied loads.

Even when a unit shuts down, normalized deviance during turnaround activities for major maintenance or process upgrade can increase catastrophic potential later. Ensure that turnaround timetables are credible. Overzealous schedules established by leaders can imply to workers that shortcuts are to be taken. This can result in a very long and possibly risky restart phase as maintenance errors or forgotten tasks are noticed upon introducing energy or chemicals to the process.

Ensure that sufficient time and resources are allocated for all process hazard evaluation and PHAs. The assigned subject matter experts should be truly expert. Assigning the newest employee to the team rather than the most experienced not only robs the team of true expertise but also sends the message that the work is not important. Team members should be committed to spend the time allotted totally dedicated to the PHA and not be pulled out of the meeting for calls, meetings, and other tasks.

6.8 EXECUTIVES SET THE TONE

Your executive, director, management, and leadership level teams, collectively and as individuals, set the tone for the level of normalized deviance your organization will accept. This level fluctuates with personnel changes, market forces, customer needs, regulatory demands, and other dynamics.

Refer to the *Executive Summary*, and evaluate how you can apply that guidance to your role as a leader, as a follower, or in such a way as to help your organization's leaders fight normalization of deviation.

6.9 SUMMARY

What can be the outcome of reducing normalized deviation to near zero? Here are two discrete anecdotal examples from CCPS member companies.

6.9.1 It Slowed Them Down. That Was a Good Thing.

A high reliability operation that produced a highly hazardous product using highly hazardous feedstocks (and produced highly hazardous byproducts) determined its operating procedures were in need of upgrade. They embarked upon a total operating procedure upgrade over two years to improve procedure accuracy and quality. The original plant was 50 years old at the time, and although modern SIS and other engineering technology upgrades had been installed, the procedures and the workers' performance had suffered. This was from the lack of organizational rigor in implementing their management of change process. MOC initiators were often checking *No* when prompted to answer the question *Do operating procedures need to be developed, deleted, or revised due to the change? Yes or No?* without first checking.

Near the end of the procedure upgrade project, the site went into a planned major site turnaround. They used this opportunity to focus on the turnaround MOC items and the recently upgraded procedures to make the startup procedures for the restart accurate and user friendly. Great effort was applied to ensure every document that their operations team would need to restart this complex plant was accurate, as detailed as needed, and matched the configuration of the changed plant.

Once the plant was up and running, the operations manager was questioned on whether or not the new procedures had any effect—good or bad—on the startup. The operations manager thought for a moment and responded.

Using the procedures and signing off each step slowed them down. But that was a good thing. In my entire time here, this startup was the first time we came up after a turnaround of this magnitude with on-specification product. It often takes days to line out the process after a startup to get our output within desired quality parameters.

The old procedures had been so out of date they were not used consistently. This was the first time in anyone's memory that workers had accurate new procedures in hand for a total plant startup. Management supported and encouraged their mandatory use throughout this critical phase of operation. Not deviating from the work processes paid off with days of high-yield, high-quality production that can never be regained once they are lost. It also paid off by showing the entire organization how to repeat this success.

6.9.2 What We Thought We Knew Was Wrong.

The production of pelletized plastics for the automotive industry requires a precise control of color and color match from batch to batch.

An internal proposal was made at a global chemical manufacturer to install online controls to improve the variation of extruded and pelletized plastics for the automotive industry. The proposal was sought due to poor current performance as measured by the amount of scrap and the pass rate of batches of pelletized plastics. This high level of deviation caused customer problems that needed to be addressed.

This manufacturer decided to install very expensive online color matching equipment to control their quality problem. However, it was very costly and the manufacturer decided to visit their joint venture in Japan before beginning the project. The Japanese site's quality control of colored pelletized plastics was reported to be excellent even without the installation of this expensive online equipment.

The Japanese facility produced material similar to that of the global manufacturer's site and had an extremely high quality pass rate with little or no scrap. The product they made in Japan had little or no variation from batch to batch, giving them a very high customer approval rating. What was the difference in this plant? The process of producing an exact color of pelletized plastic requires the preparation of a master batch. The master batch contains as many as 10 different pigments to achieve the exact color required for an automotive part. That pigment master batch is then introduced to an extruder along with the plastic material and the colored pellets are produced.

The consistency was achieved in Japan by the exact preparation of the master batch. When interviewed, the Japanese manufacturing manager stated, *When the master batch is prepared correctly, the final product will also be of the right quality.*

The Japanese workers were preparing the master batch by manually weighing very exact amounts of the various colors that are used to make up the master batch and their individual quality performance, by not deviating from the operating procedures, was perfect. There was minimum deviation from the required final color in the extruded pellets. Consequently, the global manufacturer in the joint venture scrapped their plans to install an online measuring system and concentrated on training their operators to produce a master batch that had little or no variation. A large investment was avoided and quality improved significantly.

The company was able to meet all its quality specifications without testing the final products. The process allowed zero percent testing with 100% of the quality requirements met. They achieved their consistent high quality production through procedure assurance. Strict adherence to internal work process rules was the essential piece once the procedures were validated, assured, trained upon, and followed consistently.

6.9.3 Document and Follow Your Work Processes to Succeed

You undoubtedly have similar success stories in your plant experience. When a plan is good and it is followed well, the desired results are more likely to occur. There are undoubtedly more, and more entertaining, or even tragic, stories in our industry where the plan was bad to start with, (the original deviation was allowing the bad plan to be approved) or the plan was good but deviation from it occurred over time.

The techniques and examples in this book provide industry with tools to recognize and respond to normalized deviance. It is up to you to keep this concept in mind throughout your work life, find new ways to identify and fight normalized deviation, and to promote these concepts in others and in organizational planning. However, mind the weather. It can change.

APPENDIX A – A SURVEY TO HELP IDENTIFY WARNING SIGNS OF DEVIATIONS

A group of process safety professionals derived this list for the AIChE/CCPS book *Recognizing Catastrophic Incident Warning Signs in the Process Industries* from their personal experience and through their study of historic catastrophic incidents [CCPS 2012]. Many of the warning signs represent deviations to look for or signs of already normalized deviance. Others are processes that may reveal deviation. This list can be used as a discussion tool in safety meetings or related training workshops. Some discussion guidance is provided below.

- Identify specific warning signs you have seen in your work experience.

- Discuss how your organization has addressed specific warning signs when they existed.

- Discuss how an organization can attempt to prevent a specific warning sign from appearing.

For additional guidance, as needed, please refer to the more detailed survey that is provided by CCPS [CCPS 2012].

A.1 LEADERSHIP AND CULTURE

1. Operating outside the safe operating envelope is accepted
2. Job roles and responsibilities not well defined, confusing, or unclear
3. Negative external complaints
4. Signs of worker fatigue
5. Widespread confusion between occupational safety and process safety
6. Frequent organization changes
7. Conflict between production goals and safety goals
8. Process safety budget reduced
9. Strained communications between management and workers
10. Overdue process safety action items

11. Slow management response to process safety concerns
12. A perception that management does not listen
13. A lack of trust in field supervision
14. Employee opinion surveys give negative feedback
15. Leadership behavior implies public reputation is more important than process safety
16. Conflicting job priorities
17. Everyone is too busy
18. Frequent changes in priorities
19. Conflict between workers and management concerning working conditions
20. Leaders obviously value activity-based behavior over outcome-based behavior
21. Inappropriate supervisory behavior
22. Supervisors and leaders not formally prepared for management roles
23. A poorly defined chain of command
24. Workers not aware of or not committed to standards
25. Favoritism exists in the organization
26. A high absenteeism rate
27. An employee turnover issue exists
28. Varying shift team operating practices and protocols
29. Frequent changes in ownership

A.2 TRAINING AND COMPETENCY

30. No training on possible catastrophic events and their characteristics
31. Poor training on hazards of the process operation and the materials involved
32. An ineffective or nonexistent formal training program
33. Inadequate training on facility chemical processes
34. No formal training on process safety systems
35. No competency register to indicate level of competency achieved by each worker
36. Inadequate formal training on process-specific equipment operation or maintenance
37. Frequent performance errors apparent
38. Signs of chaos during process upsets or unusual events
39. Workers unfamiliar with facility equipment or procedures
40. Frequent process upsets
41. Training sessions cancelled or postponed
42. Procedures performed with a check-the-box mentality
43. Long-term workers have not attended recent training
44. Training records not current or are incomplete
45. Poor training attendance is tolerated
46. Training materials not suitable or instructors not competent
47. Inappropriate use or overuse of computer-based training

A.3 PROCESS SAFETY INFORMATION

48. Piping and instrument diagrams do not reflect current field conditions
49. Incomplete documentation about safety systems
50. Inadequate documentation of chemical hazards
51. Low precision and accuracy of process safety information documentation other than piping and instrument diagrams
52. Material safety data sheets or equipment data sheets not current
53. Process safety information not readily available
54. Incomplete electrical/hazardous area classification drawings
55. Poor equipment labeling or tagging
56. Inconsistent drawing formats and protocols
57. Problems with document control for process safety information
58. No formal ownership established for process safety information
59. No process alarm management system

A.4 PROCEDURES

60. Procedures do not address all required equipment
61. Procedures do not maintain safe operating envelope
62. Operators appear unfamiliar with procedures or how to use them
63. A significant number of events resulting in auto initiated trips and shutdowns
64. No system to gauge whether procedures have been followed
65. Facility access procedures not consistently applied or enforced
66. Inadequate shift turnover communication
67. Poor quality shift logs
68. Failure to follow company procedures tolerated
69. Chronic problems with the work permit system
70. Inadequate or poor quality procedures
71. No system for determining which activities need written procedures
72. No established administrative procedure and style guide for writing and revising procedures

A.5 ASSET INTEGRITY

73. Operation continues when safeguards are known to be impaired
74. Overdue equipment inspections
75. Relief valve testing overdue
76. No formal maintenance program
77. A run-to-failure philosophy exists
78. Maintenance deferred until next budget cycle
79. Preventive maintenance activities reduced to save money
80. Broken or defective equipment not tagged and still in service
81. Multiple and repetitive mechanical failures
82. Corrosion and equipment deterioration evident
83. A high frequency of leaks

84. Installed equipment and hardware does not meet good engineering practices
85. Improper application of equipment and hardware allowed
86. Facility firewater used to cool process equipment
87. Alarm and instrument management not adequately addressed
88. Bypassed alarms and safety systems
89. Process is operating with out of service safety instrumented systems and no risk assessment or management of change
90. Critical safety systems not functioning properly or not tested
91. Nuisance alarms and trips
92. Inadequate practices for establishing equipment criticality
93. Working on equipment that is in service
94. Temporary or substandard repairs are prevalent
95. Inconsistent preventive maintenance implementation
96. Equipment repair records not up to date
97. Chronic problems with the maintenance planning system
98. No formal process to manage equipment deficiencies
99. Maintenance jobs not adequately closed out

A.6 ANALYZING RISK AND MANAGING CHANGE

100. Weak process hazard analysis practices
101. Out-of-service emergency stand-by systems
102. Poor process hazard analysis action item follow-up
103. Management of change system only used for major changes
104. Backlog of incomplete management of change items
105. Excessive delay in closing management of change action items to completion
106. Organizational changes not subjected to management of change
107. Frequent changes or disruptions in operating plan
108. Risk assessments conducted to support decisions already made
109. A sense that we always do it this way
110. Management unwilling to consider change
111. Management of change item review and approval lacks structure and rigor
112. Failure to recognize operations deviations and initiate management of change
113. Original facility design used for current modifications
114. Temporary changes made permanent without management of change
115. Operating creep exists
116. Process hazard analysis revalidations not performed or are inadequate
117. Instruments bypassed without adequate management of change
118. Little or no corporate guidance on acceptable risk ranking methods
119. Risk registry is poorly prepared, non-existent, or unavailable
120. No baseline risk profile for a facility
121. Security protocols not consistently enforced

A.7 AUDITS

122. Repeat findings occur in subsequent audits
123. Audits often lack field verification
124. Findings from previous audits are still open
125. Audits are not reviewed with management
126. Inspections or audits result in significant findings
127. Regulatory fines and citations have been received
128. Negative external complaints are common
129. Audits seem focused on good news
130. Audit reports are not communicated to all affected people
131. Corporate process safety management guidance does not match a site's culture and resources

A.8 LEARNING FROM EXPERIENCE

132. Failure to learn from previous incidents
133. Frequent leaks or spills
134. Frequent process upsets or off-specification product
135. High contractor incident rates
136. Abnormal instrument readings not recorded or investigated
137. Equipment failures widespread and frequent
138. Incident trend reports only reflect injuries or significant incidents
139. Minor incidents are not reported
140. Failure to report near misses and substandard conditions
141. Superficial incident investigations result in improper findings
142. Incident reports downplay impact
143. Environmental performance does not meet regulations or company targets
144. Incident trends and patterns apparent but not well tracked or analyzed
145. Frequent activation of safety systems

A.9 PHYSICAL WARNING SIGNS

146. Worker or community complaints of unusual odors
147. Equipment or structures show physical damage
148. Equipment vibration outside acceptable ranges
149. Obvious leaks and spills
150. Dust build up on flat surfaces and in buildings
151. Inconsistent or incorrect use of personal protective equipment
152. Missing or defective safety equipment
153. Uncontrolled traffic movement within the facility
154. Open and uncontrolled sources of ignition
155. Project trailers located close to process facilities
156. Plugged sewers and drainage systems
157. Poor housekeeping accepted by workers and management
158. Permanent and temporary working platforms not protected or monitored

159. Open electrical panels and conduits
160. Condensation apparent on inner walls and ceilings of process buildings
161. Loose bolts and unsecured equipment components

APPENDIX B – JOB AND TASK ANALYSIS

Source: *Recognizing Catastrophic Incident Warning Signs* [CCPS 2012].

All technical training benefits from using the instructional systems design (ISD) model for determining and delivering the specific knowledge, skills, and attitudes a worker requires to succeed in their job. Deviation is avoided by mastery. The following describes a very basic approach. Compare this with how your facility designs and implements training.

B.1 JOB AND TASK ANALYSIS AND THE INSTRUCTIONAL SYSTEMS DESIGN MODEL

The analysis phase of the instructional systems design (ISD) model consists of a job and task analysis based upon the equipment, operations, tools, and materials to be used as well as the knowledge and skills required for each position. Most important in this phase is the selection of the performance and learning objectives each employee must master to be successful in their job.

B.2 BASIC STEPS FOR A JOB AND TASK ANALYSIS

1. First, collect information about the job position from various sources such as:
   - Job descriptions
   - Standard operating procedures, maintenance procedures, emergency response plan procedures, administrative procedures, safe work practices
   - Equipment lists
   - Operating or maintenance manuals
   - Other current data such as written reports, memos, instruction sheets, P&IDs

2. Develop a list of tasks. A task can be defined as a series of steps leading to a meaningful outcome.

3. Review the list with your subject matter experts (SMEs). Subject matter experts are generally persons who can competently perform the task. This review will enable you to remove tasks that are no longer performed, add any tasks that were omitted, and correct any errors.

4. With the subject matter experts, rate each task in terms of frequency, importance, and difficulty.

5. Summarize the ratings to identify which tasks are critical, that is, which ones are most frequent, important, and difficult.

6. The following information is needed for each task:

   - What steps are taken to perform the task?
   - What standards are used in performing the task?
   - Under which conditions will the task be performed?
   - What tools, equipment, and references are used in performing the task?
   - What safety precautions does the person performing the task need?
   - What knowledge, skills, and attitudes are needed to perform the task?
   - What training does a person need to perform the task?

The design phase examines the best methods with which to impart the knowledge or develop the skills required to achieve the objectives. It allows management to lay out the most cost-effective training plan. Development is the phase in which the modules, such as process overview training or safety training modules, are physically constructed or purchased. As operating and maintenance procedures become training tools as well as job aids and hazard and process safety documentation, their revision or development could be considered as part of this phase in the ISD training model. The implementation phase represents the delivery of the training. Classroom, self-study or computer-based training are typically suitable for knowledge-related objectives. Hands-on performance training, walkthrough or simulation is most appropriate for procedure tasks.

The evaluation phase of the model exists to ensure that the organization has learned from the experience of doing the training and applies that learning to the previous phases of the model. Evaluating the worker's performance on tests or during operation may lead to revisiting the task analysis to include missed tasks or possibly redistributing tasks among the positions. This phase is a self-audit of how effectively and efficiently the training was completed.

Applying this training model results in three levels of training topics.

1. **Fundamentals**: Fundamentals include topics such as basic hazards, personal protective equipment and techniques, pressure, temperature, flow, general safe work practices, regulatory training, and common processing steps as appropriately indicated by the analysis phase.

2. **Process Overview**: The process overview includes topics related to the equipment configuration, chemical and physical changes, and special safe-work practices related to the operations, maintenance, and materials. Emphasis should be given to any new equipment and chemical hazards unit and change startup teams will encounter.

3. **Job Specific**: Job-specific topics include training on new or revised operating, safety, and maintenance procedures. It could include emergency response plan training or laboratory technician training as well if those procedures were changed.

A curriculum developed for an operator position at a facility would list

- the fundamental training the employee received upon hiring (or had completed previously),
- process overviews for each task they have performed, and
- job-specific procedures for the equipment and the continuous processing or batch instructions for each task they are expected to perform.

When a person, a team, or an organization develops a naturally high level of operational discipline (that is, when all players consistently characterize the values and attitudes that support high reliability organizations, HRO) better performance is to be expected by all workers and leaders. Some beneficial aspects in reducing the effects of normalization of deviation are listed below.

- The organization itself will generally display a higher level of operational discipline in its behaviors.
- Reducing injuries and modifying employee behavior by reinforcing safe work practices and eliminating at-risk behavior
- Reducing costs related to injuries and incidents
- Developing communications skills among all workers
- Raising overall safety awareness
- Increasing observation skills
- Developing leadership skills
- Communicating the organization's commitment to safety

Well-defined roles are essential to effectively implement any work process. Job task analysis can be performed on every position in the facility from the leadership positions to entry level jobs. This helps show the workers the company has an understanding of their roles and helps new employees succeed. Success can be defined as avoiding deviation from the work process path. Success thus demands addressing any hint of normalized deviance.

REFERENCES

Note: These references and associated internet websites (if applicable) were current at the time they were accessed during this guideline's preparation (2013-2018).

AFPM 2016	American Fuels and Petrochemicals Manufactures (AFPM), *Process Safety Performance Indicators for the Refining and Petrochemical Industries*, ANSI API RP-754, Quarterly Webinars, www.api.org/oil-and-natural-gas/health-and-safety/process-safety/rp754-webinars
AIChE 2018	American Institute of Chemical Engineers, *The AIChE Academy.* www.aiche.org/academy
Amyotte 2011	Amyotte, P. R., "Are classical process safety concepts relevant to nanotechnology applications?", *Journal of Physics: Conference Series, 2011, Vol.* 304, No. 1, 012071.
Ashforth 2003	Ashforth, D. E., and V. Anand, "The Normalization Of Corruption In Organizations," *Research in Organizational Behavior,* 2003, Vol. 25, pp. 1–52.
Bainbridge 1983	Bainbridge, L., "Ironies of Automation", *Automatica,* 2003, Vol. 19, No. 6, pp. 775-779.
Baker 2007	Baker, J. A., and F. L. Bowman, G. Erwin, S. Gorton, D. Hendershot, N. Leveson, S. Priest, I. Rosenthal, P. V. Tebo, D. A. Wiegmann, L. D. Wilson, *The Report of BP US Refineries Independent Safety Review Panel* (2007). www.bp.com/bakerpanelreport
Balsam 2003	Balsam, P. D., and A. D. Bondy, A.S., "The Negative Side Effects of Reward," *Journal of Applied Behavior Analysis,* 2003, Vol. 16, No. 3, pp. 283–296.
Banja 2010	Banja, J., 2010, "The Normalization of Deviance in Healthcare Delivery," *Business Horizons,* Mar-Apr 2010, Vol 53, Issue 2, pp. 139–148,

Belke 1998	Belke, J., "Recurring Causes of Recent Chemical Accidents," U.S. Environmental Protection Agency Chemical Emergency Preparedness and Prevention Office, 1998. pscfiles.tamu.edu/safety-alert/recurring-causes-of-recent-chemical-accidents.pdf
Bloch 2016	Bloch, K., *Rethinking Bhopal: A Definitive Guide to Investigating, Preventing, and Learning from Industrial Disasters*, IChemE, Elsevier, Amsterdam, Netherlands (2016).
Brainy 2017	Brainy Quote® 2017, www.brainyquote.com/quotes/quotes/w/warrenbenn402360.html
Brief 2001	Brief, A. P., and R. T. Buttram, J. M. Dukerich, "Collective corruption in the corporate world: Toward a process model," *Groups at work: Theory and research*, M.E. Turner (Ed.), Laurence Erlbaum Associates, Mahwah, NJ (2001).
CCBS 2018	Cambridge Center for Behavioral Studies (CCBS), "Introduction to Behavioral Safety." www.behavior.org
CCPS 2001	The Center for Chemical Process Safety (CCPS), *Revalidating Process Hazards Analysis*, AIChE/John Wiley & Sons, New York, NY (2001).
CCPS 2003	The Center for Chemical Process Safety (CCPS), *Guidelines for Investigating Chemical Process Incidents*, 2nd Edition, John Wiley & Sons, Hoboken, NJ (2003). Currently being revised.
CCPS 2006	The Center for Chemical Process Safety (CCPS), *Business Case for Process Safety, Second Edition*, American Institute of Chemical Engineers (2006; currently being revised). www.aiche.org/ccps
CCPS 2007	The Center for Chemical Process Safety (CCPS), *Guidelines for Risk Based Process Safety (RBPS)*, John Wiley & Sons, Inc., Hoboken, New Jersey (2007).
CCPS 2008a	The Center for Chemical Process Safety (CCPS), *Guidelines for the Management of Change for Process Safety*, John Wiley & Sons, Hoboken, NJ (2008).
CCPS 2008b	The Center for Chemical Process Safety (CCPS), *Incidents That Define Process Safety*, John Wiley & Sons, Hoboken, NJ (2008).

CCPS 2011	The Center for Chemical Process Safety (CCPS), *Conduct of Operations and Operational Discipline,* John Wiley & Sons, Hoboken, NJ (2011).
CCPS 2012	The Center for Chemical Process Safety (CCPS), *Recognizing Catastrophic Incident Warning Signs in the Process Industries,* John Wiley & Sons, Hoboken, NJ (2012).
CCPS 2013	The Center for Chemical Process Safety (CCPS), *Guidelines for Managing Process Safety Risks During Organization Change,* John Wiley & Sons, Hoboken, NJ (2013).
CCPS 2015a	The Center for Chemical Process Safety (CCPS), *Guidelines for Defining Process Safety Competency Requirements,* John Wiley & Sons, Hoboken, NJ (2015).
CCPS 2015b	The Center for Chemical Process Safety (CCPS), *Guidelines for Initiating Events and Independent Protection Layers for LOPA,* John Wiley & Sons, Hoboken, NJ (2015).
CCPS 2016a	The Center for Chemical Process Safety (CCPS), *Guidelines for Integrating Management Systems and Metrics to Improve Process Safety Performance,* John Wiley & Sons, Hoboken, NJ (2016).
CCPS 2016b	The Center for Chemical Process Safety (CCPS), *Guidelines for Implementing Process Safety Management Systems,* John Wiley & Sons, Hoboken, NJ (2016).
CCPS 2018a	The Center for Chemical Process Safety (CCPS), *Vision 20/20.* www.aiche.org/ccps
CCPS 2018b	The Center for Chemical Process Safety (CCPS), *Guidelines for Siting and Layout of Facilities, Second Edition,* American Institute of Chemical Engineers, John Wiley & Sons, Hoboken, NJ (2018).
CCPS 2018c	The Center for Chemical Process Safety (CCPS), *Dealing with Aging Process Facilities And Infrastructure,* American Institute of Chemical Engineers, John Wiley & Sons, Hoboken, NJ (2018).
CCPS 2018d	CCPS *Process Safety Metrics: Guide for Selecting Leading and Lagging Indicators,* Revised April 2018, aiche.org/ccps.

Cherepanov 2009	Cherepanov, V., and T. Feddersen, A. Sandroni, "Rationalization in Decision Making: Why We Don't Always Choose Our Favorite Option," *Kellogg Insight,* Kellogg School of Management at Northwestern University, Evanston, IL, July 1, 2009. insight.kellogg.northwestern.edu
Collins 2016	Collins, E. P. and B. Najafi, *"Use of Process Incident Data for Accident Sequence Precursor Analysis,"* American Institute of Chemical Engineers 2016 Spring Meeting, 12th Global Congress on Process Safety (GCPS), Houston, TX, April 11–23, 2016.
Cook 2000	Cook, R. I., "How Complex Systems Fail," *Cognitive Technologies Laboratory,* University of Chicago (2000). web.mit.edu
CSB 2007	U. S. Chemical Safety and Hazard Investigation Board, Investigation Report, *"Refinery Explosion and Fire, BP Texas City, Texas, March 23, 2005,"* Report No. 2005-04-I-TX, issued March 2007.
CSB 2009	U. S. Chemical Safety and Hazard Investigation Board, Investigation Report, *"Sugar Dust Explosion and Fire, Imperial Sugar, Port Wentworth, Georgia, February 7, 2008,"* Report No. 2008-05-I-GA, 2009.
CSB 2016	U. S. Chemical Safety and Hazard Investigation Board, Investigation Report, *"West Fertilizer Company Fire and Explosion,"* Report 2013-02-I-TX, 2016.
Dunford 2008	Dunford, R., and R. Palmer, "Organizational Change and the Importance of Embedded Assumptions," *British Journal of Management,* 2008, Vol. 19, Issue 1, pp. S20-S32.
Dunning 2014	Dunning, D., "We Are All Confident Idiots: The Trouble with Ignorance is That it Feels So Much Like Expertise," *Pacific Standard,* Oct 27, 2014. psmag.com
EPA 2017	U.S. Environmental Protection Agency (EPA), *Accidental Release Prevention Requirements / Risk Management Programs,* 40 CFR Part 68, Clean Air Act, Section 112 (r)(7), Washington, D. C. (1996, Amended 2017). www.epa.gov/rmp
Ermann 2002	Ermann, D. M., and R. J. Lundman, *Corporate and Governmental Deviance: Problems of Organizational Behavior in Contemporary Society,* Oxford University Press, New York, NY (2002).

Gerstein 2008	Gerstein, M., *Flirting with Disaster: Why Accidents are Rarely Accidental,* Union Square Press, New York, NY (2008).
Hotten 2015	Hotten, R., "Volkswagen: The Scandal Explained," *BBC News,* 10 December, 2015. www.bbc.com
Kletz 1993	Kletz, T. A., *Lessons from Disaster: How Organizations Have No Memory and Accidents Recur,* Gulf Professional Publishing, Houston, TX (1993).
Kletz 2009	Kletz, T. A., *What Went Wrong? Case Histories of Process Plant Disasters and How They Could Have Been Avoided,* 5th Edition, Butterworth-Heinemann/IChemE (July 2009).
Klein 2016	Klein, J. A., "The ChE as Sherlock Holmes: Investigating Process Incidents," *Chemical Engineering Progress,* 112(10):28-34 (2016).
Klein 2017	Klein, J. A., and B. K. Vaughen, *Process Safety: Key Concepts and Practical Approaches,* CRCPress (2017). www.crcpress.com
Kruger 1999	Kruger, J., and D. Dunning, "Unskilled and Unaware of it: How Difficulties in Recognizing One's Own Incompetence Lead to Inflated Self-Assessments," *Journal of Personality and Social Psychology,* December 1999, Vol. 77, No. 6, pp. 1121-1134.
Lawrence 2007	Lawrence, T. B., and S. L. Robinson, "Ain't Misbehavin: Workplace Deviance as Organizational Resistance," *Journal of Management,* June 2007, Vol. 33 No. 3, pp. 378-394.
Milner 2016	Milner, G., *Pinpoint: How GPS is Changing Technology, Culture, and Our Minds,* W. W. Norton & Company, New York, NY (2016).
Mlodinow 2012	Mlodinow, L., *Subliminal: How Your Unconscious Mind Rules Your Behavior,* Pantheon, New York, N.Y. (2012).
NASA 1986	National Aeronautics and Space Administration (NASA), *Report of the Presidential Commission on the Space Shuttle Challenger Accident,* Washington, D.C., June 1986. history.nasa.gov/rogersrep/genindex.htm
NRC 2000	U.S. Nuclear Regulatory Commission (NRC), *Technical Basis and Implementation Guidelines for A Human Event Analysis Technique (ATHEANA),* NUREG-1624, Rev. 1, U.S. Nuclear Regulatory Commission, Washington, D.C. (2000). www.nrc.gov

NRT 1996	The National Response Team (NRT), U.S., *The National Response Team's Integrated Contingency Plan Guidance* www.osha.gov/laws-regs/federalregister/1996-06-05-0
OSHA 1992	U.S. Occupational Safety and Health Administration (OSHA), 29 CFR 1910.119. *Process Safety Management of Highly Hazardous Chemicals,* Washington, D.C. (1992). www.osha.gov
OSHA 1998	U.S. Occupational Safety and Health Administration (OSHA), 29 CFR 1910.109. *Explosives and blasting agents,* Washington, D.C. (1998). www.osha.gov
OSHA 2018	U.S. Occupational Safety and Health Administration (OSHA), "Suggestions to Prepare for a Successful Stand-Down," www.osha.gov/StopFallsStandDown
Phung 2018	Phung, A., "Behavioral Finance: Key Concept-Overreaction and Availability Bias," *Investopedia, 2018.* www.investopedia.com
Pink 2009	Pink, D., *Drive: The Surprising Truth About What Motivates Us,* Riverhead Hardcover, New York, NY (2009).
Pinto 2008	Pinto, J., and L. Carrie, K. P. Frits, "Corrupt Organizations or Organizations of Corrupt Individuals? Two Organizational-Level Corruption Phenomena," *Academy of Management Review,* 2008, Vol. 33, No. 3, pp. 685-709.
Reason 1990	Reason, J., *Human Error,* Cambridge University Press, Cambridge, U.K. (1990).
van Stralen 2018	van Stralen, D., *High Reliability Organizing, Managing the Unexpected.* www.high-reliability.org
Vaughan 1983	Vaughan, D., *Controlling Unlawful Organizational Behavior: Social Structure and Corporate Misconduct,* University of Chicago Press, Chicago, IL (1983).
Vaughan 1996	Vaughan, D., *The Challenger Launch Decision: Risky Technology, Culture, and Deviance at NASA,* University of Chicago Press, Chicago, IL (1996).
Vaughan 1999	Vaughan, D. 1999, "The Dark Side of Organizations: Mistakes, Misconduct, and Disaster," *Annual Review of Sociology,* 1999, Vol. 25, pp. 271–305.

Vaughan 2004	Vaughan, D., *Organizational Rituals of Risk and Error: Organizational Encounters with Risk,* Hunter, B., and M. Power, Editors, Cambridge University Press, Cambridge, U.K., pp. 33–66.
Vaughan 2008	Vaughan, D., "Interview with Diane Vaughan at Columbia During AMCF," May 2008. www.consultingnewsline.com
Vaughen 2011	Vaughen, B. K., and T. Muschara, "A Case Study: Combining Incident Investigation Approaches to Identify System-Related Root Causes," *Process Safety Progress,* 30:372–376 (2011).
Weick 2007	Weick, K., and K. Sutcliffe, Managing the Unexpected: Resilient Performance in an Age of Uncertainty, Jossey-Bass a Wiley Brand, Hoboken, NJ (2007).

INDEX